DOUBLE HYENAS
AND
LAZARUS BIRDS

DOUBLE HYENAS
AND
LAZARUS BIRDS

A SIDEWAYS LOOK AT THE PACIFIC OCEAN
AND EVERYTHING IN IT

CHARLES HOOD

HEYDAY
Berkeley, California

Copyright © 2025 by Charles Hood

Library of Congress Cataloging-in-Publication Data
Names: Hood, Charles, 1959- author.
Title: Double hyenas and Lazarus birds : a sideways look at the Pacific
    Ocean and everything in it / Charles Hood.
Description: Berkeley, California : Heyday, 2025.
Identifiers: LCCN 2024044970 (print) | LCCN 2024044971 (ebook) | ISBN
    9781597146661 (paperback) | ISBN 9781597146678 (epub)
Subjects: LCSH: Natural history—Pacific Area. | Pacific Area—Description
    and travel. | Hood, Charles, 1959—Travel—Pacific Area.
Classification: LCC QH198.A1 H66 2025 (print) | LCC QH198.A1 (ebook) |
    DDC 598.09182/3—dc23/eng/20241209
LC record available at https://lccn.loc.gov/2024044970
LC ebook record available at https://lccn.loc.gov/2024044971

Cover Art: Charles Hood
Cover Design: Archie Ferguson
Interior Design/Typesetting: Diahann Sturge-Campbell

Published by Heyday
P.O. Box 9145, Berkeley, California 94709
(510) 549-3564
heydaybooks.com

Printed in East Peoria, Illinois, by Versa Press, Inc.

10 9 8 7 6 5 4 3 2 1

It is undone business
I speak of, this morning,
with the sea
stretching out
from my feet
—Charles Olson, *The Maximus Poems*

Contents

Author's Note

Families are hard to write about. I agree with Janet Malcolm when she says that "the past is a country that issues no visas. We can only enter it illegally." My version of who said (and did) what may differ from the versions remembered by other family members; that is normal enough, and all I can say is that the strongest rope is the one that has the most strands. I hope that my memories braid well with the memories of the others around me.

At least I got most of the nature right. There are sixty-six species of birds in this book, though only a few are named. I saw them fly past and made a mental note, but I usually don't trouble the reader with them. There are a dozen kinds of fish and eight species of whales. There are ninety-some-odd quotations in this book, almost all of them accurate. To the surprise of my friends, there are only two Bob Dylan references, and both are brief.

Somebody once said the writer Brendan Constantine was a true poet. The reviewer said that's because he knows all about things he knows nothing about. So too here. Many wise people have helped me work on these excursions and essays; their names are listed in the back. Errors and infelicities remain my own, unless some passage in particular really bugs you, in which case that part was added by my editor and I had nothing to do with it.

GOOD WATER, BAD WATER

I would like water better if it were not always trying to kill me.

The first time I died I was five. It was Easter weekend, clear and hot, and in the middle of a normal Saturday afternoon I drowned in a swimming pool in Las Vegas. How I fell in I don't remember, but once I accepted the fact that I was dead, I could enjoy the greeny-blue color of the light at the bottom of the pool. There was no thrashing or panic: like two tiny pink ballast tanks my lungs filled with water, and as I passed out, the colors seemed peaceful, nice. Even now, blue is one of my favorite colors.

Hours passed. What urgent beast dragged me back to the surface? It was a large fish, I think, or my tall, strong mother, or a body belonging to one of the many strange faces crowded over me, blocking the sun as I lay calm and dead on the warm cement. Eventually there was a trip to the hospital, though now that I think about it, I am not sure why. One is either drowned permanently or else only drowned for part of the day, and there is nothing a priest or ER doctor can do to change either state of being.

The next time was in high school. I had been gifted a two-week climbing course by my parents, held near Sequoia National Park. Given what a klutzy weakling I was at most sports, when it came to climbing, I was surprisingly competent, even talented. I liked everything about it: the secret lingo—"On belay? Belay on!"—and the clickity sounds the carabiners made, and learning the knots, and hearing the stories of the Yosemite hardmen, men like Yvon Chouinard and Royal Robbins and Warren Harding, men who had done bold leads on big walls and had not been stopped by running out of water or having all their pitons pop out during a screaming fall.

We were sleeping in tents and peeing in the bushes and not taking showers (because there were no showers), and since two weeks is a long time to climb without a break, it was agreed we would take a day off and go to a place called Lake Hume. I still couldn't swim—any time I tried to learn, water tried to drown me again—but everybody else was a yes vote, so I thought I could take a book and watch from a warm rock while the others did laps or jerked off in the seaweed or did whatever it is that high-octane male teenagers do in cold water.

Short story shorter, once we got to the lake, I was grabbed from behind by some boys from the adjacent Christian camp (who apparently thought I was somebody named George), thrown in, and held under. They were trying to prank one of their own but got the wrong sap.

Ever since Las Vegas, I had (and still have) a pathological fear of water, and one thing about runty guys, we can fight like honey badgers when we have to. I got back to shore not by swimming or calling for help but by smashing and hitting and grabbing. I kept trying to get ahold of something that would squish or detach, assuming that any injuries would slow them down enough to let me use their bodies as ladder rungs to get back to shore.

Finally, I panic-thrashed my way back to land. I looked at my hands. There was blood under the nails. Mine or theirs? I had cuts on my legs and knees, so I was not sure. I must have swallowed a lot of water, and, legs shaking, I leaned against a Jeffrey pine, throwing up in the ceanothus.

Once upon a time, Earth had no water at all. And then, somehow, it did—and not just a little, but pretty soon, oceans and oceans of the stuff. In the Quran, God sends water down from Heaven. It is up there, presumably abundant, and with Allah's divine concern, now it is down here. I also like another theory, the one that guessed water arrived on comets and meteors during something geophysicists call the Late Heavy Bombardment, even though more recent work disappointingly suggests you can get a planet's worth of water just from early magma interacting with atmospheric hydrogen. Whatever alchemy was involved, it worked. According to the United States Geological Survey, adding up all the moisture on, in, and above Earth produces a total of 332,500,000 cubic miles of water. Only a little bit hangs out in muddy trickles like the Amazon or the Yangtze or the two Hearst Castle swimming pools; most of that volume is in the ocean. And so that means if we borrowed all the ocean water on Earth for an hour and magically patty-caked it into a sphere, it would create a ball of water larger than Pluto.

Pluto is welcome to it as far as I am concerned, yet as one says about a bad relationship, it's complicated, since I am totally and utterly devoted to seabirds. I like their weird names, I like how they fly like magic superjets, I like how we know so little that almost any trip I take, I have the possibility of contributing to science through casual observation. It's a big term, "seabirds," since they range from sparrow sized to albatrosses with wingspans four feet wider than Shaquille O'Neal's widest, stretchiest reach. There are over a hundred main seabird species, plus one hundred kinds of gull and tern. There are seabird species whose world populations number in the tens of millions (Wilson's storm-petrel), and there are some species whose tally barely comes to 250 birds. That one is called Bryan's shearwater, and it nests in the Bonin Islands, south of Japan. It is the world's smallest shearwater and was only described in 2011. Are there other, tinier ones waiting to be found besides that one? Probably.

Seabirds are found in every ocean of the globe. And therein sits the conflict, since the Almighty has decided to keep the seabirds in the sea, and the sea, dagnabbit, is also where the planet keeps the water. To experience seabirds, I have to go to them, which means water and boats and fear, lots of fear. All water is insatiable; no matter how many shipwrecks you feed it, no matter how many naval tragedies and drowned sailors, it always wants more. Ocean, I name you squall doom and hunger comb, salt burn and plunder mouth. I name you starlight's coffin. I name you enemy.

Even so, I reluctantly admit, saltwater really is as interesting as promised. In parts of the ocean, water circulates so slowly, it has been out of contact with the atmosphere for hundreds of years. It will rise up and remix eventually, but not in our lifetimes. Black smokers are jets of iron sulfide shooting out of the ocean floor at seven hundred degrees Fahrenheit, creating exotic ecologies we will next encounter on an ice-clad moon of Saturn. There is an ocean animal, the Longman's beaked whale, that is one of water's long-standing mysteries. Until 2003, the only evidence for its existence came from two weathered skulls: one found on an Australian beach in 1882, and the other on the floor of a Somali fertilizer factory in 1955. It is not true that we know the moon better than the deep ocean (since we know a lot about the ocean, all in all), but it is true that both continue to surprise us. For me, I like any ecology that can be home to the coelacanth fish, last known from when the comet took out the dinosaurs and presumed to be extinct until 1938, when one was found off the coast of South Africa. So taken as a whole, they are jazzy places, these oceans. Too bad they have so much water in them.

Birders call their ocean excursions pelagic trips. "Pelagic" in that sense means away from shore and out onto the open sea. Marine biologists use the same word differently, so to them pelagic is a distinct boundary under the surface. In their field, modifiers get tacked on: mesopelagic, bathypelagic, abyssopelagic, hadalpelagic. In that taxonomy, each band of the ocean is deeper than the one above. If you could core sample the water the way geologists drill rock, each section would have its own color, salinity, temperature. Do the lowest levels resent the higher strata, the ones always pressing down, down? Maybe they prefer being cold and still, a kind of basement water that is unperturbed by the churning chaos of the wind-driven surface.

Style and duration of pelagic trips—again now meaning bird study—vary. Some only go out for a few hours. On a short trip, you go out, look around, come right back. It is not much longer than the director's cut of a movie. Other times the expedition lasts for days and days consecutively, and one time I sought seabirds for a full month, when I signed up for a transect that started in New Zealand and went up through the Central Pacific—the Pacific of my father's service in World War II, and the Pacific of the atom bomb—all the way to Japan.

In making these trips, wonders began to accumulate like medals on a general's chest: Here is my service ribbon for when I crossed over the Mariana Trench, the gash in the planet that drops down more than six thousand feet deeper than Mount Everest is tall. This silver bar commemorates seeing a squid that glows in the dark. Waterspouts get their own pip, as do each of the Seven Seas.

There have been the usual problems and detours along the way. I have, for example, been lost at night kayaking in the South Pacific, and I have continued to try—and invariably failed—to learn to swim, failed despite multiple lessons from multiple kindly, well-intentioned instructors. "Tenth time is the charm," I told myself the last time I tried (and failed) to master the strange condition of being in the water and not drowning. Sometimes I get lucky and find ways to avoid water while still getting hits on my pelagic wish list. That happens when seabirds come to me, brought by storm or error to the desert, which is where I live when I am not on boats.

One of these times was the summer of 2023. Tropical Storm Hilary passed over my house in the dark, and by dawn petrels from Mexico were dropping out of the sky like black hailstones. In one lake near my house, there were more wedge-rumped storm-petrels in one place than had ever been counted before by all the bird-watching boats from all the ports of California put together. The other inland birders and I were high-fiving. We couldn't believe our luck.

My kids can tell you about driving home from church and how I stopped the car in the middle of an intersection, shouting "*jaeger, jaeger, jaeger,*" and running down the street. A jaeger is a deep-chested, falcon-winged ocean bird that chases lesser birds with the velocity of a TOPGUN graduate taking down a Cessna 150. They do this because they want to harry the victim—usually a tern or small gull—until it pukes up a fish, which the jaeger then scarfs up. A folk name for it is jiddy hawk. To see a jaeger in the desert was improbable but not impossible. As I knew, there were previous records. But there are three similar species. Which kind was this one? I chased after it, trying to get a better view.

The engine was still on. The turn signal was still blinking—I had been in the middle of turning left out of the bank parking lot when the bird appeared. My driver's door was open. The jaeger was going twenty or thirty miles an hour, so I was never going to catch up. That didn't stop me from trying. What about my family, back in the car?

Oh, *them.* Right. I jogged back to apologize.

Water in small doses is domestic, passive. It will take the shape of any glass you pour it into. The problem is that water grows bolder in the aggregate. If it is pushed back by the strong walls of the dam, it wants to burst out and flood all the downstream towns. In the ocean, if water hits up against the shore, it wants to scour the beach and carry the sand far away. It will take the beach-front houses too, as soon as it can reach them. Lifeguards agree: if you are playing on the beach, you may think you are on land, but the sea knows you are not. They are called "sneaker waves" for a reason.

Not so many years ago all oceans were dangerous and unknown, at least to the people of Europe (being ignorant of the civilizations of Oceania), but even so, at the same time oceans were sexy and full of possibility. Think about Johnny Depp as Captain Jack Sparrow: how good we all looked then, with dreadlocks and kohl and a slim bone awl daggered through our red headbands. His look has artistic precedent, since Mr. Depp is both reenacting and yet updating the N. C. Wyeth illustrations for *Kidnapped* and *Treasure Island.* Those books remind us that in the old days, oceans carried you from the mundane to the exotic—all you had to do was walk up a gangway and set off. My father joined the navy in 1940, over a year before Pearl Harbor, just as the United States was poised on the abyss of an immense and unstoppable war. I once asked him why—why did he enlist at such a precarious time?

"Because I was hungry," he told me. "Because I needed a job." Pause. "And because I was tired of North Carolina. I wanted to see the world."

To see the world—he was quoting recruiting posters but also ocean liner ads and movies like *Road to Singapore,* a prewar, love-conquers-all travelogue starring Bing Crosby, Bob Hope, and Dorothy Lamour. As buxom and bejeweled dancers dip and sway, the movie's trailer promises an hour and a half with the "sun-bronzed sirens of the South Seas." Smooching a girl at home was hard work: everybody knows your business, and in my father's case, he was the preacher's son with an older sister, six older brothers, and an entire congregation watching over him. There was an unspoken sexual promise inherent in enlisting; docking in foreign ports meant one could participate in things that were immoral and illicit, and nobody at home would ever find out. Sexual adventure was not only possible, it was expected. Officially of course this only included heteronormative options. "What ain't we got?" ask the swishy, crop top–wearing, laundry-doing Seabees in the musical *South Pacific*. "We ain't got dames!" In actual practice, long before Tennessee Williams moved to Key West, sailors helped make it the hippest, queerest port in North America—which is probably why he came in the first place.

The sea now is a danger zone of nanoplastic and drone attacks, and the conveyor belt for the shame and promise of world commerce. Even science is no longer joyful; when a new way of generating air was discovered, a natural event that comes directly from stones on the seabed, it was labeled "dark oxygen," as if it were a malignant process. If shipping makes the news, it usually is linked to air pollution or Houthi rockets, or worse, pilot error. When the container ship *Ever Given* got stuck crosswise in the Suez Canal, I heard that one writer's entire print run was on board. If that was true, I hope somebody's tenure didn't rest on timely publication. While the legal teams fenced and parried, unloading the contested ship was delayed for months, and in the end, some cargo took a year to arrive.

"The sea defines us, connects us, separates us," as Philip Hoare says. Compared to the oceans of Melville and N. C. Wyeth, the world's seas are smaller now, anyway—smaller and more litigated. The Suez-blocking *Ever Given* was "Japanese-owned, Taiwanese-operated,

German-managed, Panamanian-flagged, and Indian-manned," according to the *Guardian*. It was going from Malaysia to Rotterdam at the time of the incident, which was its second accident in two years. (It previously had smashed into a ferry in Hamburg.) What country was its insurance company based in, and afterward, which nation hosted the servers that relayed the settlement's demands and counter-demands? We are all our brothers' keepers now. The earth does not even stop at the earth, since during the blockage, you could download images from satellites that were tracking the two-hundred-ship tail-back from the serene and unregulated silence of outer space.

In the military, what is now called "being deployed" (since humans now are "assets," like drones or cartons of cigarettes) used to be called "shipping out." In World War II, the Germans shipped out to go to Norway or Tunisia; the Tommies went to Gibraltar or Burma; the Americans crossed by sea to end up on Attu or Guam or Iwo Jima.

Even if I am going out for only the day, a pelagic trip still feels like a kind of shipping out. I have my haversack of rations and my well-cleaned, cloth-wrapped weaponry, and I may not be wearing a helmet, but I still have an appropriate hat in birding-approved khaki. There will be maps and radar domes and a route of attack: modern trips vector current lines and cut grids across seamounts—nobody looks for birds or whales randomly.

To name the colors of the ocean, start by listing pewter and bronze, then move through cobalt, cool black, Aegean, zinc and onyx, blood turquoise, Delft, smalt, the coiled allure of oil-sheen blue, the Chrysler blue of boneyard engines, cadet blue, boat-wake blue, and go all the way to the ghastly blue of the dead dog I saw in 1971 on a beach in Crescent City, California. Washed up under a dull, ceramic sky, the carcass was swollen, hairless, cave eyed. I was eleven, and I dared my brother Fred to touch it. We both ran away. The oceans may have been mapped already, radar sounded, satellite surveilled, but I still want to see them for myself. I follow other tertiary explorers in this, since I live when I live and need to fill in the maps, the lists, in ways still available to me. ("If you wanted to be first / You live in the wrong time," writes poet Katharine Coles about Antarctica.) After a multiyear truce, water and I have started to dance the dance again.

In my pursuit of seas and seabirds, one species I have wanted to see is the Markham's storm-petrel. The best place to look for them is in the Humboldt Current, twenty miles off the coast of Peru. The shore itself marks the start of the Atacama Desert. The Atacama barely gets half an inch of rain a year, and in some places the only annual precipitation is fog condensing directly on the bare ground. Atlas keepers label it the driest desert in the world, after Antarctica.

The Atacama is bare rock and negative space, stretching in all directions. Offshore, though, currents and upwelling provide rich feeding grounds. That is where the storm-petrels are. The catch (there is always a catch) is that no commercial trips go out there, so I had to charter a fishing boat. That sounds rustic, even romantic, but basically this was a large wooden rowboat with an outboard motor lashed on the stern and a plastic barrel for fuel. No radio, no bathroom, no spare parts. There was no cabin. No deck. No maps.

For compass, we had our phones and the guide's faulty GPS unit. Life jackets? No way. There were not even any seats. To paraphrase Groucho Marx, any boat I can afford to hire is not a boat I want to go out in.

There were four of us: me, the boat driver, and a nature guide I had hired online, about whom the less said the better. After getting my bald-tired rental truck stuck in the sand on the way to the coast, he hogged the only semi-inhabitable place on the boat and edited TikTok uploads the rest of the voyage. Bird ID was up to me. Fourth voyager was my friend José Gabriel, the Nicaraguan bat biologist. When José is not home in Nicaragua he lives in Arizona, and in the mammal-spotting world, his skills are legendary. Once, when we were turning over plywood to look for snakes, a mouse darted out instead. He caught it with his bare hands, with a reaction time so fast it would have made a cobra jealous. He and I write books together and also take hard, strange trips together, though on some of the trips we edit the details before talking to our families afterward. The agreement is that I won't tell his wife what really happened if he won't tell mine. José Gabriel used to be an engineering major, and he designs all our camera traps, MacGyvering Nat Geo–quality rigs using only Tupperware, secondhand cameras, and wire from the kitchen junk drawer.

On the day of departure, it was dark and overcast. The seas that day were high—too high to be safe—and as we left the harbor, I was reminded of Stephen Crane's short story, "The Open Boat." Crane is best known for writing *The Red Badge of Courage*, but he was also a journalist. On his way to Cuba to cover the

Spanish-American war, Crane was in a shipwreck, and after thirty hours in a ten-foot lifeboat, he and a few others finally reached the Florida shore.

We were not even supposed to be out that day; the Peruvian Coast Guard had closed the ports because the conditions were too dangerous. We went anyway—it was that or forfeit our deposit. To see the birds, we went out twenty miles to a current line and poured out a slick of fish guts, the smell of which helped to draw in storm-petrels, shearwaters, skuas, white-chinned petrels, and early in the day even a few Inca terns—slim gray birds with red beaks and white moustaches flashed with yellow.

In the story, the characters row and row, always on the verge of being swamped. Crane's story starts like this: "None of them knew the color of the sky. Their eyes glanced level, and remained upon the waves that swept toward them. These waves were gray, except for the tops, which were white, and all the men knew the colors of the sea. The line between sky and water narrowed and widened, and fell and rose." The story (as did the author's real-life experience) has a grim ending, since not everybody makes it.[1]

---

1 David Foster Wallace: "When I teach school I always teach Crane's horrific 'The Open Boat,' and get all bent out of shape when the kids find the story dull or jaunty-adventurish: I want them to feel the same marrow-level dread of the oceanic I've always felt, the intuition of the sea as primordial nada, bottomless, depths inhabited by cackling tooth-studded things rising toward you at the rate a feather falls." Like anybody else besotted by metafiction, I once wrote a story told only through footnotes, but since David Foster Wallace uses them so well and so David Foster Wallace-ly, this is the only footnote you will find in this book. That sigh you hear is mine.

Birding was good, but the conditions were not. When we dropped into the trough of waves, all one could see in all directions was water. It was dark and grim, and nobody had a good time. The swell rolled and the birds whirled, and we did our best to hang on and shout out ID pointers as the birds swirled around us in a chaotic circle.

On the way back we had almost made it to land when the engine ran out of fuel, and without an engine, we could not stay pointed into the waves. You need to use either a motor or oars to keep a boat aimed into the swell, taking waves head-on, or else you turn sideways and the waves swamp you, and at that point the boat flips or sinks or breaks apart, or all three at once.

Unlike Crane's story, there were no oars. We needed the engine, and we needed it *right now*. The swell was pushing us sideways, and even if we did not swamp immediately, the swell was going to keep pushing us sideways until it dumped us onto the beach, a few miles away.

That sounds trivial, but these were ten-foot breakers, and the desert shore was an uninhabited band of cliffs. Even from miles away, you could hear the surf smashing down on the rocks with a relentless *boom, boom, boom*. The boat would overturn, that was certain, and without a life jacket—and of course we had no life jackets, not even a Styrofoam cooler lid—even a strong swimmer would get held under by the brutal surf. Even if you survived that, you would be trapped against the cliffs.

I saw that José Gabriel had his SOS beacon out. It could send a text message by way of satellite, but the issue was, send a message to whom? There was no rescue service; there was no 911 phone number.

I didn't ask, but I guessed that José was silently composing a farewell message to his wife. For me, since I have had so much practice dying in water already, I was both anxious and blasé. "So this is how it all ends," I thought, shrugging.

One of my heroes, H. W. Tilman, had led audacious, minimalist expeditions to Everest in the 1930s and had served with distinction in World Wars I and II. He died while sailing to Antarctica, aged eighty. I had hoped to approach his same octogenarian status before water caught up with me, but it seemed that I would not hit that marker. I looked around and said in my best Gatsby voice, "At least you made it this far, old sport."

In a final act of hope and desperation, tilting the fuel barrel on its side sluiced a final ounce of gasoline into the connector hose. The engine started again. With power, we could steer, and if we could steer, we could turn into the swell and claw our way back into the harbor.

Once back, it was time to consider the day's results.

Had we died? We had not.

Had we seen the Markham's storm-petrel? We had.

In the birding world, those two facts are all that matters.

This is the part of the story where a hand usually comes up from the audience. "Wouldn't a normal person . . ." followed by some kind of sensible observation about better precautions or even not having gone out in the first place. Time now to state the obvious: (a) I am not a normal person; (b) whether my listing is or is not a manifestation of OCD or any other disorder is not the point; (c) maybe I should travel with my own life jacket? (d) I have in the past, like when I once went snorkeling in Palau, but (e) if you are going to be on the road for weeks and weeks, as José and I were going to be, you have to limit your checked bag to twenty kilos, to comply with airline regulations. Forgive my vulgarity, but we already had so much shit with us, you would not believe it. That was because part of the trip was also about photographing bats, and he and I use haute couture lighting kits for this, to wrap the light and get really handsome, nonthreatening portraits of otherwise misunderstood animals. Besides the lighting gear and the spare lenses, we also had field guides and binoculars and Swiss Army knives and toilet paper and parachute cord and Covid vaccination cards and hand sanitizer and AirPods and blister packs of Ciprofloxacin. We might even have had two pairs of underpants each. There was no way I could travel around with a spare life jacket or two, just in case a chiseling fisherman tried to shortchange us by not bringing enough fuel.

During most of our trips—during the parts of the trip when we are not about to die at sea—there is a lot of slack time, like when we are driving ten hours from Lima to Nazca. Naturally one talks about the scenery and work and people we know in common, and how we can get better photos next time, and which camera bodies we most covet. After that, things begin to drift. In my case, I grow more associative and more unstructured the longer I talk. José Gabriel switches between polarities. One is that I am almost nearly interesting to be around, and the other is that my monologuing is grounds for justifiable homicide.

Here are some of the things I can remember talking about: the clarinet glissando that starts Gershwin's *Rhapsody in Blue*; Werner

Herzog; mink coats; Anglo-Saxon strong-stress prosody; the history of the Los Angeles freeway network; my perpetual search for the world's best tiramisu; and, at some point, how the "petrel" part of the name "storm-petrel" comes from Peter, Christ's disciple who walked on water during a storm. That of course reminded me that a traditional name for storm-petrels is Mother Carey's chickens, which collates several expressions at once, including a slang expression for an angry mob and the Italian phrase "*Mater Cara*" ("Dear Mother"), aka the Virgin Mary. As recently as the 1820s (I told José), sailors thought that storm-petrels laid their eggs directly on the water and brooded the young under their wings. In the nineteenth century if you found eggs on shore, like from gulls or murres, you put some in your hat to take back to the ship to eat later.

Why he puts up with me, I do not know. (I wonder that about most people.) In my case, I don't want to eat seabirds or their eggs; I just want to see them—*all of them*. It is an impossible goal, but even so, I am working on it steadily. That schlurping sound you hear is my bank account circling the drain and disappearing out of sight. Loan me your credit card and I will keep going. Sod my water fear: I want to see it all, feel it all, know it all. If my puritan ancestors espoused moderation in all things, I missed that lesson. After Bob Dylan cut a fifteen-minute version of "Highlands" for the album *Time Out of Mind*, a record company executive said, "So, Bob, do you have a shorter version of that song?" Mr. Dylan answered, "That *was* the shorter version."

I love the ambition of that. Some things can't be done in half measures or petite bites. One of the birds that José and I saw from the Peru death ship was a least storm-petrel. Sparrow sized and sooty black, it is the smallest seabird in the world. The least storm-petrel is strictly a Pacific Ocean bird, with a territory that stretches from the middle of California down to Peru, usually staying within a hundred miles of shore. It nests on islands in the Gulf of California and eats plankton. Sometimes it is put in its own genus, *Halocyptena*, a name that combines the words for "sea," "speedy," and "winged." According to the Audubon Society, the least storm-petrel flies "low over the waves with fast deep wingbeats, giving it a rather batlike look." (I might be tempted to say, "be more specific," since bats vary as much as birds do. But never mind that for now.)

When José and I saw one in Peru that grim, gray day, we were encountering it at the far bottom of its range, since that was as far south as this species ever gets. I also have seen it over the Cordell Bank, which is a marine sanctuary off the coast of Bodega Bay, north and west of San Francisco. In perfect symmetry, that particular least storm-petrel was as far north as they usually get. I love that I have gotten to experience this species at each end of the parabola. I have friends who rarely travel more than fifty miles from home, but for me, my radius is more like the swallow's, the peregrine's, the storm-petrel's. I don't mind starting the year out at one latitude and finishing the year at another meridian, half a planet away. I follow the seasons and my heart, driven by whatever strange hungers urge me forward that particular month. I like paintings of full-rigged sailing ships, and I like out-of-print nautical handbooks. I like that hour at the end of the day, after the sun has set and the water inks down into the purest, bluest black. I like tidepools and seagulls and watching dogs learn how to surf. I love it that Bob Dylan once rhymed "help" with "kelp." I like reading about rare whales, but I like seeing them even better. When water comes to take me the final time—and it *is* coming, I remain certain of that—I will try to be both worthy and ready.

Water—my friend, my enemy, my dream, my doom.

A TROPICBIRD FOR TRUDY

When I first started birdwatching, my mentor was a woman from New Jersey named Trudy. Born in 1919 after the end of the First World War, she grew up speaking German, Yiddish, Spanish, and English. (She said she could curse in eight languages.) After college Trudy enrolled in Columbia Medical School, but the other students, all men, resented her, and they sabotaged her work. The all-male teaching staff did nothing to stop them. She was a woman; the assumption was that she was going to wash out anyway—better sooner than later. It didn't help that she still went by her given name, Gertrude, which, like other Germanic, bottom-heavy names like Bertha sounded too much like the name of an artillery piece and not enough like the name of a potential wife-to-be.

The spite and harassment became too much, and she dropped out. During the Second World War, rather than become a nurse (as expected), she became a weather observer on board dirigibles. If a storm was coming, the crew went out to meet it. Did she keep a list of her dirigible-borne bird sightings? I never asked. During the war she changed her name and went by the more socially acceptable Trudy. She told me she was not a feminist; she was a survivalist.

Trudy was broad shouldered and broad bosomed but only five feet tall, defiant in red lipstick and a dyed-black men's cut. She did not believe in seatbelts (she said they cut into her chest) and told me the policeman had not been born who could make her wear one. After the war she moved to Southern California. When I asked her about how legal or less than legal it was to bird the Irvine Ranch property in the 1950s, Trudy was the one who taught me the expression, "If you're not trespassing, you're not birdwatching."

In the field, Trudy kept a .22 handgun wrapped in flannel in the glovebox of her jeep and used a folding entrenching tool to dig herself out of ruts. These army shovels have a small, spade-shaped blade and a short, sturdy handle. The blade pivots at a hinge to fold back over itself for storage. She had bought hers at an army surplus store, one of those now-gone, back-in-the-day emporiums whose underlit aisles offered everything from parachute cord to brass knuckles to waterproof ammo cans. Before REI and Big Five, it's where outdoorsy people shopped. She later gifted it to me, reminding me to sharpen all my tools once a year, shovel blades included. Worn and olive green—even the wooden handle was once painted green—it lives behind the seat. It has been with me five trucks in a row, and as instructed, I keep it sharp and rust free.

She was a patient naturalist and would go through a flock of gulls over and over with me until I got them all right. In time, even the hybrids made sense. Trudy had been taught gull ID by the legendary Guy McCaskie, though she abbreviated his name to

GM, short for the Great Man. "Good birder, but too full of himself," she told me, shrugging. She also did not like French people and always called seasickness *mal de mer*, as if blaming the French for inventing it.

Trudy's favorite bird was an ocean species few of us have heard of: the red-billed tropicbird. It is one of a trio of related birds called "elegant" by *Birds of the World*, which adds that they are blessed with "long central tail feathers that stream behind them, lending their passage a grace that few birds can match." It goes on to note that even though tropicbirds are ungainly on land, in flight "these birds are aerial acrobats, whether they are dodging the thievery of frigatebirds or performing the remarkable 'back-pedaling' of a courtship display." Do we watch birds in envy for all that we are not? Few of my own courtship displays are ever elegant.

On the West Coast, all tropicbirds are gift birds: you don't go to them, they come to you. A few drift into California airspace each year at the end of summer, arriving in onesies and twosies from nesting sites in Mexico. There is no one place to expect them—you just have to go out to sea and start hoping.

The red-billed tropicbird is three feet wide wingtip to wingtip, and also three feet long from the serrated red beak to the end of its floating tail. Sitting on the water, tropicbirds keep the tail streamers lifted up, dainty and dry. Rich Stallcup in *Ocean Birds of the Nearshore Pacific* says in flight, "tropicbirds seem to have only one gear—high, and full throttle." To hunt they plunge-dive after fish and squid, or else chase down flying fish on the wing. One Caribbean term is *flèche-en-cul*, which Googles translates without comment as "arrow in ass." About a thousand pairs of arrow-free red-billed tropicbirds nest in the Gulf of California. You can also see them in the Caribbean and the Galápagos, but Trudy always said that was cheating. She liked the challenge of trying to find one in her home circle, and preferably Orange County waters, not some foreign location like San Diego or Ventura.

There are two other tropicbird species worldwide, both similar to the red-billed. All have tail streamers. One called the white-tailed tropicbird is satin white with black racing stripes on the wings. Trudy was a kind and patient mentor, but because she had been birding for so long before I was born, she had accumulated a list of astounding sightings. One example of this dates from 1964, when California's first (and so far only) white-tailed tropicbird was seen near Upper Newport Bay. The bird was present for a month. I quote now from Dunn and Garrett's *Birds of Southern California*: "The bird was attracted to remote-controlled glider airplanes which hobbyists flew from the bluffs of the bay. The misguided bird actively courted and even tried to copulate with these gliders." As the song reminds us, if you can't be with the one you love, love the one you're with.

Since 1964, there have been no other California records of white-tailed tropicbird. Trudy saw it, Guy McCaskie saw it, and I am sure many other people saw it, but for those of us born in later times, that species gets filed under the heading "far away and long ago."

It turns out that 1964 was a good year for inland seabirds. At the same time the lovesick tropicbird was harassing model planes, a flock of blue-footed boobies took up residency in what is now called

Bonelli Park, near the Los Angeles Fairgrounds. Trudy tried not to gloat when she told me she had seen those too.

I've had mixed luck with red-billed tropicbirds myself, if luck can be the right umbrella term for a final score of missing them on fifty-two California boat trips and seeing them on just three. You can go to their home ranges and have better odds, but for most birders, Trudy is right. Seeing them in one's home territory (usually for a state or county list) carries a particular thrill. For the white-tailed tropicbird, since no more have shown up in California, to add them to my life list I had to go to the Dry Tortugas, below the Florida Keys. I also have seen them on the Big Island of Hawaii, where they nest inside the caldera at Hawai'i Volcanoes National Park.

To encounter the third and final kind, red-tailed tropicbird, requires researcher status, or it does in North America. It shows up in the rare bird logs only because there are survey boats quartering the ocean, right out at the very edge of California's two-hundred-mile territorial limit. The rest of us, limited to regular boats going out from regular ports, have no chance.

In order to see my first-ever red-tailed tropicbird, I had to go to the South Pacific. "Cheater," I can hear Trudy crowing now. My excuse was that the birds were incidental—I was working on a book about the atom bomb plane, the *Enola Gay*, and I ended up retracing its route from Wendover, its training base in the Utah desert, to each of the islands it stopped at before launching the Hiroshima strike. The Hiroshima mission took off from Tinian, but to get there I first had to connect by way of Saipan, which was a site of one of my father's own pieces of the war. The two of us echoed each other, both looking up, both of us thinking and worrying and holding binoculars, squinting up at invisible shapes crossing that troubled the sky.

During this same time when I was in the South Pacific trying to die in kayaks and leaky boats, Trudy had moved from Orange County to the inland town of Hemet, in Riverside County, to take care of an ailing father. He had plumbing trouble she said, explaining that his aim in the bathroom was so bad, rather than let him make a mess, she told him to pee in a bucket. A neighborhood gent was courting Trudy around then, and he helped with caretaker duties. I never learned his real name, since in her weekly letters she only called him Swabbie, as in, "good for swabbing the decks." (With her father's shaky aim, it seems there was a lot of mopping up.) Trudy and Swabbie had met in church. All her life she was a devout Episcopalian, or as she put it, using the old joke, "a card-carrying member of Catholic Lite."

At her request, I sent her postcards from my trips, along with sketches of animals and birds. She wanted to hear about all of it, even the dark times. One especially grim story that I shared took place on the island of Saipan, in the Northern Marianas. History remains visible and present throughout the Pacific, and there also cemeteries and war memorials on most islands. You can still find intact bunkers from the war or come across the ruins of crashed planes.

In 1944 my father's ship supplied covering fire for the US Marines as they tried to take Saipan from fierce defenders. It was brutal fighting. Trapped and brainwashed, not only did the Japanese soldiers commit suicide rather than surrender, but the Japanese civilians did too, leaping to their deaths from the island's cliffs even as US interpreters used megaphones to beg them not to do it.

You can go to the cliffs today, as I forced myself to do, trying to make sense of a time when even women holding their babies would leap rather than surrender. Horror overlapped with beauty during my visit, since there are palm trees at the battle sites now, and ferns and small orange butterflies. Overhead, pairs of red-tailed tropicbirds floated in the blue and immaculate sky like kites at the end of long, invisible strings. These were new arrivals to Saipan. I was told by my guide, a schoolteacher named Samuel, that tropicbirds had not nested here before the war. They only came afterward, and it was widely understood that the birds were not birds—they were the souls of the dead children.

This was explained to me casually, soberly, plainly. Samuel was not trying to put one over on the newbie from the mainland, nor was he speaking in metaphors or symbols. The birds were not *like* children, they *were* the children, and everybody on the island knew it.

And so they were. Once he told me, I saw them that way too—and now I pass it on to you. War in the large, abstract sense may never end, but specific wars, plural and lowercase, do end, and we all have to figure out how to find hope and meaning in the world that is left over, afterward.

Tropicbirds inhabit contradictory chambers in my heart. Trudy has passed now, as have most other veterans of that war, my father included. To the extent that I associate the bird with their memories, I find tropicbirds coated with a very fine, nearly invisible sheen of resentment. That expression made popular by Tom Brokaw, "the Greatest Generation," referencing the self-sacrifice and "can-do" spirit of the Depression and World War II generations, is both true and false. They *were* great—except for how much racism they tolerated, and except for how, after the war, the men promptly told all the women to quit their jobs and get their tushes back in the kitchen, and except for the refusal to talk about problems out loud, problems ranging from alcoholism to DDT to child abuse. They were great, except for all the times they were not.

I feel guilty for saying that out loud.

And I feel guilty for being thankful that I was never stuck in the snow in the Battle of the Bulge nor pinned down on the beach at Tarawa. My life will never have to add those experiences to the catalog of memories I will have to process later, memories and fears that will come to me, unwelcome and persistent, in the middle of the night. I may have other nighttime fears, but not those ones.

On Tinian Island, the *Enola Gay*'s final base, they have tropicbirds too. There also is a postcard-ready beach, white sand and perfect coral, overlooked by coconut palms and the concrete pillboxes of the Japanese defenders. If mixed well and reinforced with rebar, concrete lasts a long time; the bunkers are still there, little changed, even after so many years. In the water, rusting at the tide line, are the leftover axles of an American tank, a piece of hardware that was blown up on July 24, 1944. You can see that same tank in the black-and-white battle photos and match those documents to the ruins visible today. Somebody died, *right here.*

Somebody very much like my father.

Somebody very much like me.

The birds of course have nothing to do with this. They are children or they are spirits or they are animals, though if so, according to the aesthetics of beauty most of us agree to share, they are fabulous animals. Seeing them, it is impossible to be indifferent. One strange thing is how poorly they fit existing taxonomic lineages. All our assumptions turned out to be wrong. Ornithologists want to place them correctly on the tree of life, yet it turns out, tropicbirds are not terns. They are not pelicans. They are not boobies or frigatebirds or petrels. They may be related to two singular jungle species: the sunbittern of South American rainforests and the kagu of New Caledonia, a large island in the South Pacific. The sunbittern, according to eBird, is "one of the most dazzling of all the world's birds; the intricate yellow, red, and black pattern on the spread wings is truly otherworldly." The kagu lives half a world away and is a flightless blue chicken from the mountain forests of New Caledonia. It has red eyes and looks like a demented night-heron. I have seen them both, and neither time did I think, "Oh look, here comes a terrestrial tropicbird." But if the DNA says they are sisters from another mister, then so be it. The systematics of who is related to what relies on an arcane alchemy that is, as Trudy would tell me, far above my paygrade.

John James Audubon painted tropicbirds, creating a plate that remains in heavy rotation on his top-ten playlist. Melville includes tropicbirds in *Moby-Dick* but only as small flecks of white dabbed into the larger canvas. Attention rarely lingers. In Melville's time tropicbirds were sometimes called bos'n birds, because to the deckhands, often piped to their duties, the birds' whistled vocalizations sounded like a bosun's pipe. According to one guidebook, at sea the red-billed tropicbird emits "occasional piercing screams." Yes, but why?—I would like to imagine they do it because they're bored, or perhaps they really like giving sailors the heebie-jeebies. Tropicbirds never show up in movies, but they are popular on postage stamps. When I die (which is a finish line I feel closer to some days more than others), I would not mind coming back as a tropicbird, since it seems like they enjoy flying and they also appear to be tidy eaters, which would be a pleasant change from my current sloppy habits.

These are birds of distance—seen *at* a distance, and distant in the sense that they are distant from the places "we" inhabit (whoever "we" are): Des Moines, Paris, Berkeley, London, Trona, Hemet. No tropicbirds in New York City. No tropicbirds in the Bible. *Birds of the World* explains that "they breed on remote, isolated oceanic islands, often in very inaccessible places." This makes sense. If I were giving life advice to aliens and other exotic species, I would suggest breeding as distantly from humans as possible.

I exempt myself from this recommendation. The tropicbirds can be around *me*—I won't hurt them. In fact, I am one of their biggest fans. I just want to paint their picture in my journal, and type up a brief report of the encounter, and seal it in a letter-sized envelope, and wrap the envelope in clear plastic, and leave it with some flowers and perhaps a small flag on top of Trudy's grave, Riverside National Cemetery, Van Buren Boulevard, east of where the Pacific Plate meets the North American Plate, planet Earth, Solar System, middle of the Orion Spur, Milky Way Galaxy, a short 13.7 billion years away from the first start of so much ocean and memory.

SEA LEVEL DOESN'T EXIST

The lowest point in North America is the Badwater Basin in Death Valley National Park. Two hundred and eighty feet above the parking lot, high on a desolate, rocky cliff, there is a Park Service sign declaring "Sea Level." It reminds viewers how far below grade they are standing.

The sign, of course, lies.

Every map does too, agreeing with all the others that there is one mark on the shoreline, a line that wraps all around the world, one distinct place called "sea level." Above this mark, that is where the land starts. Beneath it, that part is owned by water. Ah, more lies.

I think about sea level often, particularly when in a cartographical mood. My garage holds boxes and boxes of paper maps, left over from the time in my life when you needed a paper map in order to go hiking or explore a new city. Like books, you could never have enough maps, and I had them not only for places I had been but for places I might go to next, like the Simien Mountains in Ethiopia or the Harding Icefield in Kenai Fjords National Park. These days my archive is partially depleted: I've cut up 20 percent of my maps for collages and art projects, and a few maps got used as kindling during wet days in the backcountry.

In reviewing the remaining stock, I see that most of my USGS topo maps (the "regular" hiking maps produced by United States Geological Survey, originally sold in flat paper sheets) state that "datum is mean sea level." I like the found poetry of that statement, though I remember misreading it in high school as datum is mean *at* sea level, and I thought, "So just take datum to higher ground and maybe it will be more pleasant." What USGS is explaining is that a thousand high tides and a thousand low tides meet in the statistical middle. How tide height was measured in a presatellite, predigital era was by using float gauges that were installed in stilling wells that dampened the wave action and, through a system of wires and pulleys, translated the vertical movement of the float to a pen that traced out a line on a paper chart attached to a rotating drum. It was like an earthquake seismograph but for water. With enough stilling wells on enough coasts, meaningful averages could accumulate.

When I lived at Newport Beach, I loved to watch the tide come in at the Wedge, where a south swell hitting at the right angle bounces off the harbor jetty and recombines with new, incoming waves to create thirty-foot breaks. *Ka-boom*: it is steep and mean, giving bodyboarders short, powerful rides before it sucks them out in an intense backwash. (Experienced wave thrashers only need apply: "If in doubt, don't go out.") This site is at its best in summer and fall, when hurricane-driven storms come slanting up from the equator. Each set is different. There is no "typical" or "average." Because of the interference pattern of one wave interacting with another, a lifeguard told me the only math that explains the tidal dynamic is chaos theory. Expanding out to the rest of the ocean, the best one can say is that there are a lot of sea levels, plural, but nowhere is there ever just the one. At any given moment, water the world over bulges higher one place and dips lower at another, more antipodal place. It never stops shifting.

To explain why this happens, we start with tides, and to start with tides, we start in outer space. The sun and moon each pull on the earth with their individual gravities, only they do so from two directions at once. Their vectoring gravities, tugging from their respective corners of the sky, deform the earth's oceans. As the earth rotates (which it does at a thousand miles an hour), and as these two outside bodies tug from always-changing angles, the earth's oceans get bent sideways. Piles of water lurch across the globe, carrying their bulges with them.

Because of all this spinning and sloshing and spilling and tugging, most (though not all) coasts have tides that go from low to high and back again roughly every twelve hours. These oscillations can generate large swings: in the Bay of Fundy, tides charge up and down forty feet per session. Cook Inlet, in Alaska, experiences a tidal range of thirty feet—enough to inundate a three-story house. Water never just sits there, level and calm, doing nothing—except it does, sometimes, in places on the planet's surface called "amphidromic points."

"Amphidromic" is a fancy word that means that there are places that end up being tidally neutral because of how the nearby currents bend. These places might have offshore currents (even quite strong ones), but they don't have tides. They do, however, connect up with places that do have tides, and there are complicated maps that chart out how the pressure points and the slack zones interact to create what can be, on some coasts, very steep and chaotic tides. If it is null at one place, that just means water will surge at another.

There are more complications than the tides. According to NOAA—the National Oceanic and Atmospheric Administration— "in areas of the planet where gravitational forces are stronger, the mean sea level will be higher. In areas where the earth's gravitational forces are weaker, the mean sea level will be lower." The stronger versus less strong spots are not spaced out symmetrically. A gravity map of Earth is as swirly as a tie-dyed T-shirt. Besides high and low tides and besides gravitational variation, there are the surface waves themselves, rising and falling, endlessly, endlessly. The differences in the surface of the sea generated by the currents mean that the Atlantic Ocean is three feet lower on the north side of the Gulf Stream than it is farther south.

And just to make the idea of sea level more dubious still, atmospheric pressure changes daily. Due to the inverse barometer effect, a low-pressure weather system allows water levels to rise, since the lid of planetary air presses down less firmly than usual onto the surface of the sea. Storm surge includes that pressure easement plus all the water that has been piled up downwind ahead of the storm. The piled-up, wind-driven water added on top of the higher base level creates a storm surge that overtops seawalls and deposits ships on lawns and leaves cars stranded in parking lots. For an hour or a day or a week, beachfront community X is now below sea level.

All of this means that sea level is a goal and never an achieved state—or it is a state that is not realized for longer than a few minutes, as the tide rises up to air quote "sea level" and then keeps inching past it, or else drops below the invisible line for hours at a time, such as during a minus tide. The barnacles clamp tight, prepared to wait for water's return.

In looking through my map collection, I am particularly drawn to representations of the Pacific (so much gorgeous blue), and in studying the atolls spread across the Pacific, the idea of sea level becomes more urgent than usual. I have one of my favorite maps in front of me now. It is a two-and-a-half-foot-wide map of the Pacific from 1943. This National Geographic Society product was used by thousands and thousands of American families as they followed the battles of World War II. The map is mum about its definition of sea level, but it does explain that its international boundaries date from "September 1, 1939, the day Germany invaded Poland." To convey the immense scale of the Pacific, the map lists both flying distances and sailing duration in nautical miles. Air travel was still so new that the mapmakers hesitated over the unfamiliar word, "airline," and finally spelled it as a hyphenated compound: "air-line."

On this map, everything rises up from zero. All mountains are measured from high tide, looking up. Yet an opposite perspective exists, one that privileges looking down. That is because nautical charts (unlike land-centric maps) need to ensure that ships do not run aground. A few feet either way matters a lot to a ship coming over a sandbar and into a harbor, and so nautical maps indicate sea level by measuring from the average *low* tide, on the assumption that it's better to have too much water rather than too little. High tide is numerically off limits.

The exception is when nautical charts indicate the gap between the surface of the water and the bottom of a bridge. Here the map's designers follow the same logic, except they reverse it, so that on a typical nautical chart, bridge height always assumes high-tide values, never low. Once the ship arrives on site in reality, there may well be more clearance than indicated, but in a worst-case scenario, the captain knows what to expect.

Digital maps have the same problem as paper ones: Sea level is always an approximation masquerading as a universal constant. Glacial melting makes it worse. Radar altimeters in satellites now try to add coverage and reliability to this original gauge-based dataset, but these new records only reach back to the early 1990s. That is a good start, but when it comes to measuring averages of the ocean, that is not much better than the day before yesterday. Besides that, even satellites do not cover all the coasts of the world evenly. Colonial history has always skewed the goal of total tidal knowledge; the mother country (England, Germany, the US) was more likely to have more data than the colonized one (Ghana, Namibia, Guam). That led to what the *Encyclopedia of Ocean Sciences* calls "inhomogeneous spatial sampling," meaning long-term gauges were not spread out consistently. To evaluate claims about mean sea level, you thus have to know who is measuring it, and more important, where. In Great Britain, the base height on old maps is defined in terms of "Ordnance Datum Newlyn" (ODN), which is the mean level of the sea at Newlyn in Cornwall from May 1915 to April 1921. (This definition replaced an earlier Ordnance Datum Liverpool based on measurements taken in that port in 1844.)

No matter what technology is being used, the wincing truth remains that sea level is not a fact at all, but like so much of nature, it is a declaration that we prefer to think we know what it means, instead of admitting that it's something beyond our human capacity.

In questioning sea level, I am not saying that every yardstick is wrong, nor every kitchen scale. Stable units of measure exist. Somewhere in a vault in France rests the definitive kilogram, what *National Geographic* magazine calls a "lonely hunk of platinum and iridium." How much does a kilo weigh? It weighs precisely that, at least inside the vault.

Yet even that physical kilogram has been superseded by a mathematical equation based on Planck's constant, a value borrowed from quantum physics with a lot of zeroes in it. I prefer the shiny metal cylinder myself, but then again, while I don't miss rotary dial phones, I still know how to use one. I freely admit I am not a trustworthy judge of the modern world.

Semi-stable fact or fictive cultural construct, I find it comforting that the idea of a shared global goal works the same whether measured in metric units or imperial, along desert coasts or boreal ones, in the rain or during the night or under the strange antilight of a total solar eclipse. They may start as different base units, but there still is sea level in Antarctica and sea level in London (held back, at times, by the Thames Barrier), as well as sea level lapping around the well-guarded walls of Marin County's San Quentin Penitentiary. Sea level may be a myth, but it is a persistent one, since sea level is one of the few things we have all agreed to accept at face value. Everybody uses it; nobody questions it; it remains as enduring as ever.

Sea level serves one final function: it tells the sea, "Keep Out." If sea level marks the topmost edge of the ocean, beyond which it must not travel, it also delineates the start of the nonwater half of the planet—our part, the air and land part. The idea of sea level is the shore's way of warning us about the ocean, saying, "There it is. Can you see how much of it there is? For God's sake, leave it alone."

THE HALF-LIFE OF SALT

too much of a good thing

To catch a bird, put salt on its tail—my grandmother taught me that. Interesting stuff, salt. You can clean a scorched pan with it or use it to lift wine stains out of the carpet. Humans are finely tuned salt-using machines: not enough salt, our neurons can't fire; too much salt, we seize up. We perpetually ride the razor's edge. If you live in or near water, salt is a problem, since on average, seawater is three and a half times saltier than the interior of a vertebrate body. If you're a six-foot-long adult sea turtle, the same as if you are a six-foot-tall person, you can't survive ingesting as much salt as seawater delivers, not without doing something to push it out as fast as it comes in.

Several methods to do this all overlap. In her book *How the Ocean Works*, Helen Czerski describes how a leatherback turtle swims slowly, feeding on jellyfish as it goes. "Every minute or two," Czerski says, "a dark, pulsing silhouette emerges from the [dim water], a messy cascade of orange tendrils hanging from a colorless dome." With a twist of the turtle's flippers, "One snap and a puff of debris is all that remains."

The turtle ingests a lot of salt during this transaction, because "a jellyfish is really just a small bucketful of ocean masquerading as life." Less than 1 percent of its mass is digestible as food, so a large sea turtle like a leatherback is sucking in so much seawater with each jellyfish that if the day's salt intake were extracted, dried, and weighed, it would fill a twenty-pound sack. To get around this, the sea turtle weeps. Extensive salt glands in its head remove the salt and extrude it out through the animal's tear ducts. "Leatherback tears," Czerski notes, "are thick and viscous and almost twice as salty as the ocean."

This process is a high production throughput: the typical big turtle weeps two gallons an hour. At the same time, salt and other waste is streaming out of its backside. As a sea turtle swims and eats, we are told that "huge plumes of very liquid feces emerge from the other end." It makes me want to order up some custom stickers. Every time I see a cute pink sea turtle sticker on a car in the mall parking lot, I will slap on a second sticker showing a dark plume emerging from the animal's rear end.

Seabirds face the same salty dilemma. They may be gone from land for months, even years, and a nesting island may have no freshwater anyway. They have to drink seawater and bathe in seawater and eat fish whose bodies concentrate seawater. It is more total sodium than their kidneys can process, and so marine birds have glands nestled in a depression in the skull above the eye socket. Bird expert Peter Ryan explains that "these salt glands strip sodium and chlorine ions from the blood and excrete a highly concentrated salt solution down ducts leading into the nostrils. This explains the drip of liquid often seen in the tips of seabird bills—they really do have a runny nose." The nostrils themselves can be simple external slits (such as on an albatross) or a more elaborate double-chambered, top-of-bill tube that gives seabirds the collective name of tubenoses.

"honey, let me be your salty dog"

The ocean is salty because the land is salty. Other than the first few hundred million years when the planet was getting started, the salinity of seawater has remained surprisingly constant. Mineral-laden sediment sinks to the seafloor and coalesces into rock, entombing salt with it, but then each rainy season the world's rivers deliver tons and tons of replacement sediment, so the input and output (including the water lost to evaporation) more or less balances out.

Experienced sailors used to be called old salts, and all sailors supposedly use salty language. Despite his years at sea, my father never cursed, revealing a churchy restraint I would do well to emulate. These days to be salty is to be angry or sharp tongued; to be more peeved than outright angry is to be lowkey salty. In the black-and-white era of the *Andy Griffith Show*, a salty dog was a potential suitor, as can be heard in rollicking bluegrass licks on YouTube, which preserves these and other pieces of my youth. Typical lyrics: "Standin' on the corner with the lowdown blues / A great big hole in the bottom of my shoes / Honey, let me be your salty dog."

seated below the salt

Humans enact our own version of the periodic table simply by being alive. We are made up of this amount of cobalt, this much bile, this percentage of gold, a handful of lithium and copper, a tiny dot of magnesium. Add sorrow to taste. Some of us have after-market add-ons—what is the specific gravity of tattoo ink? I had an accident once that included surgery and rehab and the *will I ever walk again?* conversation. The plates and rods that hold my right leg together must change my mineralogy, since sometimes they set off the metal detector at the airport. I am salt and light, but I am also a line of scars and $70,000 of surgical titanium.

Salt as status: At the medieval table, to eat dinner below the salt meant you were out of favor—salt was expensive and so was kept close to the important and the powerful. If you were seated below the salt, time to take note. Things were getting precarious.

Salt as family history: When I was growing up, salt was my mother's most trusted spice. Pepper was a suspicious afterthought, added reluctantly and mostly to appease my father. From scrambled eggs to biscuits to pork chops, salt was as essential as the skillet itself. My mother was brand loyal and only used Morton, which has always come in a round box with a blue label and a girl holding an umbrella. It is on my own shelf today. The hinged metal spout remains unchanged.

We are smarter about our bodies than our parents' generation was, and their parents before them. Many of my friends are vegan or vegan adjacent. Some meditate or do yoga. Most can quote poetry and often quite a lot of it. Me too, but I still like going to In-N-Out Burger—*shhhh*, please don't tell anybody. It is my private shame. But if I order my fries well done, they are cooked longer and saltier than usual, so they taste like growing up.

My mother cooked the way she did because she had not been taught a better way. In fact, she had not been taught at all—I gather there was not a lot of love or continuity going on in her girlhood. In elementary school, she was perceived by a school nurse as being underweight and received an allotment of medicinal graham crackers every day at recess, which she was instructed to eat to fatten her up. A plump child is a happy child, that was the goal. Something was going on at home though that graham crackers could not fix. Her own mother—my Nana—had a "drinking problem" as people used to say, though this was kept secret as best as it could be. "Most of all, you've got to hide it from the kids," sing Simon and Garfunkel about the troubled housewife, Mrs. Robinson. Hide things, keep them secret: this is the American way.

I wonder how many of my vegan, meditating, seemingly liberated friends have vices they can't share? One friend obsessively but very secretly collects Hello Kitty memorabilia; another friend has a surprisingly active Grindr account. I am slightly shocked when I learn that this or that friend still smokes. (David Hockney: "It used to be you couldn't be gay. Now you're not allowed to smoke. It's always something.") Not many people know this, but when I have writer's block, I have a set of Hot Wheels cars that I take out and drive around the desk. It doesn't help me write any better, but it is a pleasant way to kill time until I can break for lunch.

salt and money, money and salt

In the strange way that fate works, salt saves birds, or more precisely, salt ponds do. From Indigenous times until now, salt in California was mined in the simplest, most timeless way possible: dike up shallow ponds in a sheltered bay, fill them with seawater, let sun and wind evaporate the water, and then bag up the results. Environmental journalist Rosanna Xia explains why this mattered: "California's earliest oil refineries, chemical plants, and manufacturers thrived from this local industrialization of salt. Salt, in addition to preserving and flavoring food, is a critical ingredient for mining, for canning, for bleaching paper, even for refining gasoline." Xia adds, "Companies like Dow Chemical [the company that also brought us the Vietnam War's emblematic product, napalm] took the brine from the salt ponds and, using electrolysis, figured out ways to produce caustic soda and hydrochloric acid, which led to sodium carbonate and chlorine gas." Chlorine gas was used as an early form of chemical warfare in World War I; it is pale green, heavier than air, and reportedly smells like a mix of pineapple and pepper. Both sides used it. It is fatal if inhaled but can be thwarted temporarily by covering your face with a wet bandana. If you don't have a canteen, soaking it in urine also works. After an initial trial in 1915, chlorine was replaced by deadlier gasses and then banned by the Geneva Convention.

In some parts of California, we still process brine into salt this old-fashioned way; in the San Francisco Bay, evaporation ponds crystallize five hundred thousand tons of salt each year, while San

Diego's saltworks have been in continuous production since 1871. All salt ponds were initially based on making money, and money was a good reason to let the coastal marshes be turned over for salt production. If nothing at all had been there, *something* would have been put there—a powerplant or marina or something quote unquote "useful." Nobody just leaves a bay sitting there, idling, keys in the ignition.

In fulfilling their role as commercial enterprises, salt ponds accidentally set landscape aside, keeping it reserved for later. In places such as Upper Newport Bay and Bolsa Chica wetlands, both in Orange County, once salt ended, then the state repurposed defunct duck hunting grounds and out-of-service brine ponds to create open space and bird sanctuaries. Newport Bay's salt ponds were damaged by floods in the 1960s, which ended three decades of commercial production. That was the first step of turning Newport Bay into the nature reserve it is today.

Aunt Bea, what's for supper?

We watched TV with dinner when I was growing up, a small set with a rabbit ear antenna on top, black and white, with muddy sound. The choices were the news, which is to say the Vietnam War and the Watts Riots, or reruns like *Andy Griffith*, where I learned the salty dog song, or the *I Love Lucy Show*, all of whose episodes I knew backward and forward by the time I was ten.

We said grace before starting, and after that nobody talked. You could, however, be excused once you were finished. Since I had the TV shows memorized and I didn't care about Vietnam, I learned to eat very quickly. Even now, if I am home alone, I eat every meal standing up by the sink, hand washing the dishes as I go.

In Botswana I once saw a pack of African wild dogs finish off a gazelle in less than five minutes. Other than two hooves and a smear in the grass, there was nothing left. "Hey," I thought, "if that is the audition for the pack, I'm in."

humans would make bad whales

All sea mammals cope with salt better than we do; whales can drink saltwater and humans cannot because their kidneys are that much better than ours. Seals and sea lions too. To process water more effectively, add a kidney upgrade to the list of things humans will want to work on once we get more comfortable with gene editing and body enhancements. With better internal organs, we could become our own desalination plants.

My mother taught fourth grade all her life, tracking gentrification as the property values around her Echo Park school dropped and rose, dropped and rose, and noting too how the home languages evolved from English to Spanish to Vietnamese and back to Spanish and onward to a dozen new tongues and dialects. I never attended there formally, but I spent a lot of time hanging out, and some years she let me march in the school's Halloween parade. For science lessons she had a cardboard sheet with twenty kinds of rock and mineral glued to it. Each sample was a tiny square, thick as the tip of your thumb, epoxied into a labeled grid. Nobody cared about the mica or the rhyolite, but the hard, white square of rock salt, that one we all wanted to stick our tongues out and lick. Year after year, whenever I visited Mom's room, I secretly would go into the closet and lick the rock page. There are many degrees of being sweet, but only one way of being salty; it is the ur-flavor that preempts all others. In the poem "Meditation at Lagunitas," when Robert Hass presents the urgency of sexual connection, he talks about the hunger he feels for his partner's presence, "like a thirst for salt." According to Hebraic tradition, to quote from the Book of Sirach, "The principal things for man's life are water, fire, iron, and salt."

Ceylon, which is adjacent to Siam

When I was growing up, my family liked to stick to the routine. Sundays we had bear claws and strudel for breakfast, and the Sunday paper offered the funny pages in smudgy, misregistered color. I wore a clip-on tie, and my father taught me how to shine my wingtips until they were as bright as his. The round tins had a kiwi on them, so long ago now the brand is out of business.

We checked on my grandparents after church. Reluctant to up-grade, my grandparents still kept a hand-cranked mangle on the service porch and a red brick incinerator outside. My grandfather's globe listed "Siam" for Thailand and "Ceylon" for Sri Lanka. I grew up assuming all globes were wrong; that is what globes did—they presented a half-scale map of the world, fifty years behind political reality.

While Granddaddy monologued about the international conspiracy of the Jews, Nana shuffled into the kitchen to make my sandwich. She was jowly and frizzy-haired, stout bodied under a floral muumuu, and she made little *um-hm, um-hm* noises to herself, nonstop and almost too low for the rest of us to hear. One reason I am afraid to shave my beard is that without it, I suspect most mornings I would be a dead ringer for Nana in her declining years. When I was a young man, I looked like my father during the war, thin and tense; now that I am older and paunchier, what I resemble most is an alcoholic grandmother talking to herself in an apartment in East Hollywood.

Lunch arrived each week as the same rubbery wedge of carrot, the same glass of skim milk, and the same sandwich made from stale bread and rank cheese. For lubrication, the bread had been spritzed with Parkay margarine, which my grandmother called by its Depression-era name, "oleo." In the other room, I could hear Granddaddy still carrying out his vitriolic tirade.

I see now how hard it was on all of us, Nana included. How much invisible salt do we each swim through, hour after hour? My grandmother's um-hm's were the result of years of being forced to agree with whatever she was told she had to agree with. Her own opinions were not welcome. Closer at hand, my mother would never talk about what it meant to come home from school only to find her own mother passed out drunk on the floor, shoes off, nylons crooked, drool on her cheek. Did this happen often? Nobody in the family will talk about that either, and all the witnesses are passing on, one by one. I am left with only the residue of these memories, a residue that I don't mean to spread and yet probably do, like the way I always leave a dirty fingerprint on the light switch when I come inside after working in the garage.

dinosaurs and sharks

Or rather, not dinosaurs per se, but the ancient marine giants called ichthyosaurs—the answer to how they survived in hypersaline inland seas is that they had salt-secreting glands, the same as iguanas on the Galápagos do today. Modern sharks do this too. Sharks lack the swim bladder found in bony fish, however saltwater does add buoyancy, which allows sharks to swim with less effort than they would need in freshwater systems. That means there will never be a "Lake Huron hammerhead" or a "Mississippi River black-tipped reef shark," simply because it would be too much effort to swim through so much salt-free water.

all the rooms in the palace

We all come from family, even for those of us whose family history is full of gaps and question marks. (Absence becomes presence in that case.) My grandfather had been born in Germany and somehow ended up in Wisconsin, where he wore a bowler hat and was boss of a gang of juvenile delinquents. From Wisconsin, Granddaddy kept pushing west. When did his own parents pass? When he was still young, I think, but I don't know for sure. Did he miss speaking German? Did he miss somebody's cooking, or was there a girl he left behind on the dock, both of them bawling their eyes out? When he shaved, whose lost face did he see in the mirror looking back at him? I never asked any of that. I think there was an uncle or brother purged by Stalin in the 1930s, somebody who was one of the German immigrants who had settled along the Volga River. How did that person's disappearance touch my grandfather's life? I never asked him about that either.

And my ignorance of context is true as well for Nana, my mother's mother. Born in Sweden, she was brought young to the US, but something didn't work out, so they went back, unsailing the Atlantic, and when the return to the home country didn't work out either, Nana and her family came back once more to America. In making those trips, my grandmother lost her language, her food, her climate, her friends, her dog, and her second-favorite dress. I tried to go to the church in Stockholm where she had been christened, but it is an apartment block now.

In Sweden, before the first trip to America, her father had been part of the royal cavalry assigned to the palace. And in Wisconsin, once they arrived, they were just yokels with bad accents. When my grandfather promised her a better life in California, of course she said yes.

the half-life of salt

"Having swum in the ocean," the poet Huang Fan writes, "Salt considers soup a shallow pond." Does salt resent my paltry table? When I go to my grandparents' graves, I feel little and say less. The same when I visit the graves of my parents. I see that others around me bring flowers, weep, share stories. My brother has not been to the graves even once, so I go alone, but I acknowledge how unusual that is. Other people come with siblings to gravesites, or their partners, their children. Maybe if Forest Lawn allowed it, I would bring my dogs? I used to rock climb, and I also was interested in military history; back then, I "collected" sightings of rare tanks. More recent compulsions—chasing seabirds, chartering helicopters, going to hard, new places—are more or less versions of my earlier compulsions, except increased in cost and duration. While in theory these new pursuits are less harmful than traditional addictions, they are not any less expensive or isolating. Does the chemistry inside my head, my heart, contain a different ratio of vinegar to Karo than is typical? Hello world, come on now, make me your salty dog.

Somewhere there is a village where people say "I love you" to one another easily and readily, and every time they say it, they mean it. Just one more trip and maybe this time I will find my way there—it is getting closer now, so close I can taste it, like I can taste the salt on my lips at the end of a pelagic trip, and like I can still taste, after all these years, the intense hit of pure pleasure when I licked the rocks glued to the warped cardboard of the fourth-grade science poster, and my brain lit up as I smiled to myself, murmuring, "Ah yes, *salt*."

SUNRISE WITH SEA SNAKES

If I were going to write a murder mystery, I would kill the victim with a sea snake bite. If it were a James Bond movie, there would be an entire lagoon brimming with sea snakes.

I say that because everything about them is terrific. Sea snakes—also called sea kraits—breathe air, swim in saltwater, hunt in packs (sometimes), live in the Pacific and Indian oceans (usually), and can bite, though mostly in self-defense, unless you are a goby or eel or prawn, in which case the bite was probably intentional. They can dive three hundred feet down and sleep underwater. They have one lung, longer than a land snake's and with different musculature, to process air better.

Sea snakes are often banded yellow and black, though in some species the contrasting stripes can be blue instead of black. Sometimes the belly is solid yellow and the back all dark. These patterns may attract prey, and at the same time they wave a warning flag, since sea snake venom is potent and lethal. According to a website intended to help doctors with a diagnosis in the field, after a person is bitten by a sea krait, "paralysis, dysphagia [inability to swallow], muscle spasm, respiratory arrest, and dysarthria [inability to talk] can occur, and the most common cause of death in sea snake poisoning is respiratory arrest due to diaphragm paralysis, or [else] drowning secondary to skeletal muscle paralysis."

In other words, don't get bitten.

We usually think of sea snakes as tropical (if we think of them at all), but I was at a meeting in Los Angeles when Greg Pauly, herpetologist at the L.A. Natural History Museum, got a call and had to leave. There was a sea snake washed up in Newport Beach, and he was going to retrieve it as a museum specimen.

Greg later spoke to reporters about the snake's significance. As temperatures change, so does the distribution of animals. "Oceans are warming and the species that respond to that change will be those that are the most mobile," Pauly told the *Los Angeles Times*. "So the big question is this: Are sea snakes swimming off the coast of Southern California the new normal?"

I don't know about the new normal part, being still perplexed by the old normal, but the idea of being killed by such a gorgeous animal—as opposed to leaking my life away in the Covid ward of a hospital or dying from my combined childhood exposure to asbestos, lead paint, smog, pork sausage, and secondhand smoke—pleases me muchly. Sea snakes are so photogenic *National Geographic* editors put one on the cover of their 2023 "pictures of the year" issue.

Living in the sea has its own risks. We would expect sea snakes to get barnacles the same way that whales, wharfs, or cargo ships do, and in theory they could, and yet they don't, because they shed their skin even more often than land snakes do. Frequent renewal keeps them clean and sleek. To scrape off the old skin, they rub along the seafloor or else tie themselves up like a boa constrictor killing a rat and push out of their old skin that way. It leaves the old skin floating but knotted, a puzzle to beachcombers.

To mate, the male sea snake inserts one of his penises—he has two—into a female's cloaca while they swim underwater. They stay entwined for many minutes until the female decides she has had enough and releases the male snake. Do female sea snakes have a clitoris? Inquiring minds want to know. The details are understudied, but if female sea snakes are like female carpet pythons, female death adders, or female monitor lizards, then the answer is yes twice over, since those other reptiles have two clitorises. These nerve bundles were known previously but had been misidentified (one assumes by male herpetologists) as scent glands.

While some sea snakes lay eggs on land, in most species the female incubates the eggs inside her body. When it's time for birth, the young hatch inside of her and then exit out her cloaca and swim off. Once the young are clear of Mom, that's it—there is no care or nurturing after that. Everybody goes their separate ways.

If they successfully survive the risky juvenile stage—when the snakes are small and inexperienced and eaten by everything—then an adult can live ten years. Their nostrils are higher up on the head than a land snake's, and in most sea snakes the eyes are raised as well. Maybe in a million years they will have blowholes and swim like dolphins. For now the tail is compressed vertically like a ship's rudder, and some sea snakes even have modified their belly scales to arrive at a midline peak, forming a simple keel. An organ under the tongue concentrates salt, then discards it.

My mother was afraid of snakes, so much so she could not even watch them on television. Almost all snakes can swim, but I am not sure she knew that; when she found out that sea snakes could wash up on a beach—intruding into what should otherwise be a safe zone for humans—that struck her as simply too much. Somehow the universe was not playing fair if it put snakes in the ocean. What next, she complained—snakes coming right out of the toilet? Sea snakes like to forage along current lines, where one layer of ocean pushes up to another, the few feet of concentrated water between them trapping wood, plants, foam, small fish. Some locations encourage these currents more than others. One place in Costa Rica like that is called *Bahía de las Culebras*. The name dates back to the Spanish conquest and means "Bay of the Snakes." Mother would not approve.

My hunger for the thrill of the new ties in with my need for travel, and so in the map of my life I could put in a pushpin for each place I have seen a sea snake. Only half a dozen so far, and I remember each one, with room on the map for the ones yet to come. And here, these pins will go into the map at each of my manta ray sites, while here and here, these pins mark witness to a whale shark's languid wake. The black pins are for the times I have been far from home and sad; red for active volcanoes; white for any time somebody told me what they dreamed when I did not care to hear. Statues and/or murals of Komodo dragons at these pins, and real Komodos on each side of this island here. I said something stupid here, this pin, and wished I had been nicer, this long row of pins, contiguous from here to here. A dolphin breached in the moonlight, this pin, and phosphoresce trailed behind our boat, these pins, a rainbow arc of them, each pin smaller than the last, dwindling to the very edge of the map.

I can call them each to mind, and they are so vivid, these encounters, that there's a term for how my brain works, as a helpful friend explained to me: "hyperphantasia," or "extremely vivid mental imagery." Before he sent a link, I had always assumed my way of processing was universal and hence was puzzled by—and impatient with—people who did not seem to follow my way of thinking and describing. It felt as if they were being forgetful on purpose, or in some way deliberately thick-witted. I am sure I was just as perplexing and difficult to them.

What does a snake see when it sees whatever it sees? Color, pattern, food, not-food. The sea snake smells a thousand different kinds of water, then moves on, encountering a thousand kinds more, and another thousand, and another. It must think about us, "What stupid dolts, so ponderous and slow." (Sinuous, humans are not.) Snake and human fear the other, instinctually and quickly; serpents and humans each worry that contact will be fatal to us both.

I really, really liked finally seeing a sea snake in the wild, even though all I can say about it is that it looked cool. The way evolution works, if there were going to be land snakes, *of course* there would sea snakes, and when we get to the off-world oceans, the ones on moons like Enceladus, they will have sea snakes there too, or else something even better, stranger. It would be worth living to be two hundred years old, to be around to catalog all the new ocean things, once we get up there to see them.

Art does its best to show the way. *Sunrise with Sea Monsters*, 1845, is the title of a painting by Mr. Joseph M. W. Turner, Royal Academy. Which kinds of sea monsters he had in mind remains unexplained. There may be two fish in it—or there may not be. It is an agitated abstract yellow painting, smudgy in a modern way and probably not finished, though it can be hard to be sure. Too bad dead people can't get a five-minute timeout to come back up to check their email and answer a few questions: Hey Turner, what does all this yellow mean? In the painting, the sunrise is veiled (and amplified) by a swirling mass of—? Fog? Irradiated poison gas? Piscine flatulence? Whatever is happening, it is not a reassuring tableau. Tate Britain, the painting's current custodian, explains that "despite its bright tonality, this work may relate to Turner's frequent depictions of the sea as dark place."

Other Turner works carry titles like *A Disaster at Sea*, *Rough Sea with Wreckage*, and *Dutch Boats in a Gale*. He saw the ocean for what it was: beautiful and deadly. As his work grew more unconventional, he was seen not as a visionary but as somebody enduring a fit of madness. He was not concerned about these perceptions. In reference to *Snow Storm: Steam-Boat off a Harbour's Mouth*, Turner explained, "I did not paint it to be understood, but [because] I wished to show what such a scene was like." William Blake had been marginal all his life, so his artistic detours attracted little comment. In comparison, the public and famous Turner offended almost everybody by deviating from market conventions. His art became as unpalatable as his personality. The usual websites give summaries like this: "intensely private, eccentric, and reclusive, Turner was [ahem, clearing of throat] a controversial figure."

Maybe all great artists are a bit batso. In Turner's case, we don't know much about the interior man. The exterior one, not much more. He had blue eyes and a Cockney accent. He was short. He never married any of the women he was partnered to, though he gave his last name to his children. He was made a full member of the Royal Academy before he was thirty, which was and still is a land speed record. Hired to sketch Oxford High Street for an engraving, Turner was accurate down to the glazing on the windows and the caps on the dons. Besides paying the contracted fee, the publisher was so pleased he gave Turner a case of sausages as a bonus. In seeking commissions, Turner started with the landed gentry and then followed the money, new or old. His patrons included a horse dealer, a coach maker, a whaling tycoon, a textile magnate, and a brewer. When he sold a picture, he charged extra for the frame. He may have had cataracts. At the close of his life, Turner had to have all his teeth pulled, and at the very end, lived on milk and rum. He died beside his final partner, Sophia Booth, in her rented house, under an assumed name. Apocryphal last words: "The sun is God."

Turner kept his favorite paintings for himself and then bequeathed his collection to the nation. Critic John Ruskin helped manage his paintings and sketchbooks after Turner passed. The problem was one of decorum, since by the time of his death, tastes had changed. Turner was born in Convent Garden, the red-light district of London at the time, and lived mostly in the Georgian era, when casual sex was casually accepted and 25 percent of all births happened out of wedlock. He died fifteen years into the reign of Queen Victoria, and by then some of his sketchbooks were no longer drawing room appropriate.

To make sure the Turner legacy was preserved, John Ruskin claimed the pornographic sketches in Turner's journals may have been done by somebody else—perhaps somebody French—or if they were Turner's, then they must have been created during brief periods of mental instability. It sounded plausible, since according to the tastes of the time, only a damaged, prurient mind would find sex interesting, plus Turner's mother had died in the Incurable Ward of Bedlam, so obviously instability ran in the family.

In any case, Ruskin said there were only a few such questionable pages, and that he had burned them. He hadn't, but he said what was needed to be said to save the bequest. J. M. W. Turner is buried in St. Paul's Cathedral, in a tomb alongside those of Lord Nelson, the hero of the Battle of Trafalgar; Sir Joshua Reynolds, first president of the Royal Academy; Lawrence of Arabia (bust only; the body is elsewhere); and Sir Christopher Wren, who built the place, finishing it in 1710.

Brusque, impatient, secretive, imperious, cheap—Turner's biographers have called him many things, and few are nice. Apparently he would answer a too-obvious question with a grunt, or not even answer at all. Other times he was blunt, even caustic. If we want to label categories of behavior, these days we all have access to online tools and are free to use expressions such as "on the spectrum." It is easy to suggest that a pattern of behavior indicates that a person who is supposedly out of compliance with social norms must be exhibiting an "ism." It doesn't have to be autism itself; there are other options too, all at our quick-to-Google fingertips.

I've had this kind of instant diagnosis applied to me, and often by people who should know better. I am going to speak on behalf of private and cantankerous William Turner and say that he and I are better diagnosed by poetry, not a bullet list of putative deviations. In particular I am thinking of the poet Li-Young Lee, who says that given the world's beauty, is it "any wonder I've lived most of my life / insomniac by night / and distracted by eternity all day?" There is so much to see, feel, sing, dance, paint. There are over sixty species of sea snakes for example, and in honor of them I am going to invent a new word: "ophidiophilia," not the fear of snakes, but the love of them. I don't want a sea snake as a pet (they are better off left in nature), and I don't even intend to see every species myself. I just exult that they exist at all, and revel in the color and grace that my hyperphantastic memory holds, as bright and as dangerous in my mind as they are in the sea. I like too the strange ways that nature and culture intersect, though nature often comes out the worst for it. In 1856, William Muir of Cockburn Street, Edinburgh, designed and sewed a pair of men's slippers out of golden sea snake leather and yellow quilted silk. You can see them today at the V&A in London. The species Mr. Muir used is native to the Timor Sea, between Australia and Indonesia; how the leather arrived in Scotland, I do not know.

I wish I could paint, and if so, I would do a grand series: *Sunrise with Sea Snakes*, followed by *The Sun Is God* and then *Enceladus*,

Mermaids of the Eternal Ice. My former teacher, the poet Louise Glück, says that beauty initially leaves the mind "mesmerized or stunned." That is followed by "other sensations, none of them articulate." First, she says, comes a rush of excitement, "succeeded by a feeling of arrival, of completeness, and, with the new completeness, insatiability."

I know the feeling; beauty enters me with equal speed inside a gallery and the symphony hall as it does when I am kayaking over the coral reef, and it arrives irrespective of fear, fatigue, hunger, or annoyance. All I know is that I want more. Who I am at the moment and what I would prefer to be doing no longer matters. The same for others, or it seems. Crossing the Alps, Turner could not draw fast enough; he would lean out the coach window, filling journal pages as fast as he could. There is an Odyssean story (perhaps true) that he had himself tied to the top of a mast during a storm at sea. Was he, at that point, a danger to himself and others? I hope so.

There are times when being alive is even better than sex, and I do not say that lightly. There are sixty kinds of sea snakes—yes, I brought that up before, but come on, *sixty.* And so many snakes, and yet that is only the start of an inventory of all our cabinets of wonder. A meteorology group has an atlas that identifies over one hundred "species" of clouds. New kinds of clouds continue to be discovered. Here, let me buy a round of drinks for the namers of clouds. One hundred kinds, and they are not done? Right effing on. Other books talk about "one thousand paintings to see before you die," and we know that if we got any three art teachers together, they would disagree over those thousand choices and want to add in a thousand more, and we know too that both sets of experts would be right. (And then you and I, we could add a thousand more.) My hero Werner Herzog has made over sixty films, some of which I still need to see. Lord have mercy: so much to look at, so much to feel— I will be honest, for me it is hard to go to bed at night. I want to rest—I do not enjoy being tired, nor do I want to be the eccentric who won't follow society's expected hours—but hoosh, with all that beauty still to come, it is just hard to give up and go to bed at ten or eleven at night. My wife has to get up early for work; most nights she is in the bedroom by nine. I have tried that schedule but can't settle down. Dreams are nice and all, but Turner is better. *Sunrise with Extra Sunrise*, let that be the next project I take on. I can't paint a lick, but if I live long enough, maybe there will be time to learn. Until then, I wander around, rubbing my eyes. As Frank O'Hara has said, "And here I am, the / center of all beauty! / writing these poems! / Imagine!"

DOUBLE HYENA OLLIES A RAILING

Flying fish have escaped gravity twice. The first time was when they evolved the behaviors and body modifications that let them break free from thick, slow, predator-filled water and soar in glistening brilliance through the open air. A shark or manta ray might launch out of the water briefly, struggling to stay up for more than a second or two, but the flying fish celebrate sustained aerial journeys that cover many hundreds of feet. To stay aloft, a flying fish gains lift from the wind off the wavetops (the same as a petrel or shearwater does), and most species snap their tails side to side as well, leaving a stuttering, skipped-stone wake.

Flying fish are generally tropical and generally blue above and white silver below. Most are a foot or foot and a half long, and all species are broad winged and tall tailed. The "wings" are two (sometimes four) elongated side fins; some flying fish have a small dorsal sail, low and far back. In midflight, skimming across the ocean, the typical flying fish looks like a cross between a barracuda and a souped-up dragonfly.

Flying fish fly to avoid being eaten. Since a large predator like a tuna can outswim most everything smaller, to get away from it a flying fish accelerates in a mad burst of speed and launches clean off the surface like pulled skeet. It skim-glides through the air for one, two, or three hundred feet. From below, the sea's surface often reflects light like a mirror, so once a flying fish gets clear of the water, the chasing fish will usually not be able to track it.

To inhabit both worlds—the world above and the world beneath— flying fish bodies have to be more rigid than a typical fish's (more "torpedolike"), and they need reinforced jaws to be able to withstand impact better. Their pectoral fin-wings anchor more securely to the body than do the fins of a typical fish. But the "wings" can't become *too* elongated; the sailing part still needs to be a functioning fin underwater. The Goldilocks balance of wing versus fin was achieved early on. According to ocean researcher Peter Ryan, "fossils of flying fish, unrelated to modern species, have been found dating back 200 million years." They keep shaking the water off and inventing themselves anew.

John James Audubon, sailing from New Orleans to Liverpool in 1826, was fascinated by the flying fish he saw off the coast of Florida. In his diary, he described how schools of mahi-mahi chased them. (He called mahi-mahi fish "dolphins," a term still used in the Caribbean today.) Here's Audubon: "Dolphins move in Companies of 4 or 5 and sometimes 20 or more. [They] chase the flying fish with astonishing rapidity [and the target fish] after having avoided his Sharp pursuer in the water, emerges and goes through the air with the swiftness of an Arrow sometimes in a straight course and sometimes deviating by forming part of a Circle." Alas for them, because "frequently the whole [escape attempt] is unavailing."

Eat and be eaten, that is the law of the sea, and the mahi-mahi in turn is predated by the barracuda, and also by the sailors themselves, who caught them on handlines or gaffed them with a multi-pronged spear. One population of Cuban mahi-mahi was thought to be poisonous. To test if it is or not, cook it with money. Audubon: "A Piece [of mahi-mahi] is Boiled along with a Dollar until quite cooked when if the Piece of Silver coin is not Tarnished either black or Green, [then] the Fish is good and safe eating." Ever curious, Audubon opened up the "punch" of the mahi-mahis (the "paunch," or stomach) to see what they had been eating. He did this with flying fish too, and shearwaters and noddies, storm-petrels, and even a bottlenose dolphin the crew had harpooned and eaten.

According to a National Geographic webpage, "at least" forty species of flying fish exist. Wikipedia sees your forty and raises you two dozen, suggesting there are sixty-four species. Seabirder Peter Ryan guesses the total at "60–70 recognized species placed in seven genera." Other totals rise higher still. In his book *The Amazing World of Flyingfish*, Steve Howell says there are 150 "types," meaning forms identifiable at sea, but only "60 to 70 species known to fish biologists." This is out of thirty-two thousand species of fish in the world. As Howell says, "Surely new species remain to be discovered."

Uncertainty exists in part due to the discrepancy between a living flying fish and a dead one. Collected to be a museum specimen and pickled in a jar, each deceased flying fish looks the same as all the others. Alive, they have color, gestalt, and personality, but once dead, they are colorless, inert, and mute. They offer few clues for identity. That means to understand flying fish well enough to name them, one needs to be in the field, not in the museum. Ed Ricketts talked about this with John Steinbeck in *Log of the Sea of Cortez*, that famous mix of field notes and philosophy. "We sat on a crate of oranges," Steinbeck writes, "and thought what good men most biologists are, the tenors of the scientific world." He admitted field biologists could be "temperamental, moody, lecherous, loud-laughing, and healthy." Not everybody, though. "Once in a while one comes on the other kind— what used in the university to be called a 'dry-ball'—but such men are not really biologists. They are the embalmers of the field, the picklers who see only the preserved form of life without any of its principle."

That takes us to the flying fish's second great escape, which was to break free of language. In the field, the people who see flying fish most often are not ichthyologists but seabirders. These sunburned stalwarts spend hours on the deck of pelagic boats, logging detailed notes about anything that flies, swims, soars, or drifts past. Most whales and dolphins are covered by field guides, and almost all the seabirds too, but ichthyologists don't "do" flying fish in the field— they look at them in museums. Without field guides, to describe what they see in the field, birders have started to give each flying fish a whimsical, descriptive name. First among equals in naming is the bird guide, Steve Howell. He has spent hundreds of hours at the bow of the ship studying whales, dolphins, petrels, and flying fish, and as he says, in all that time he only got heatstroke once, and it was a mild case.

Here is a top-ten list of flying fish common names, arranged alphabetically. I should clarify that these are not accepted "in the literature" as the phrase goes, but are names given to flying fish forms/types/species by pelagic bird fanciers. They want to log what they are seeing, so they have come up with a new, observation-based taxonomy. As we learn more, the expectation is that names will evolve. For now, my favorite names include

Atlantic Necromancer
Bonin Windshield
Double Hyena
Fenestrated Naffwing
Leopardwing
Lunar Smurf
Purple Haze
Oddspot Midget
Rosy-veined Clearwing
Violaceous Rainmaker

First of all, *take that*, butterfly watchers, and second of all, compare those names to the approach shown in the *Guide to the Coastal Marine Fishes of California* (California Fish Bulletin Number 157). In that handbook, for the California flying fish, it wants us to check to be sure the specimen has "predorsal scales 47–50; midlateral scales 64–70." How about something basic and obvious, like telling us what color it is or how it flies? But then I am being unfair: if all I had was a dead flying fish, drained of color and divorced from flight style, I would have to count the scales too.

Names matter since they don't just describe the world, they create it. Even in the abstract world of so-called pure science, names reveal the biases and preconceptions of the cultures that host them. Dr. Jann Vendetti, a malacologist at the Natural History Museum of Los Angeles, surveyed five thousand Latin names for mollusks. She has published her findings in an open-access article in the *American Malacology Bulletin*. Dr. Vendetti found that when examined closely, 90 percent of the eponyms honored men while only 10 percent honored women, and while there are many historical reasons for this having to do with power and money and who mentors whom, the fact is, "gender asymmetry in molluscan eponyms likely reflects barriers to women's participation in malacology, taxonomy, and systematics." That in turn means that "recognition of this inequity should inform discussions about female representation in scientific names," and, by extension, about female representation in science more generally.

This discrepancy extends to common names, though those were not her primary focus. With the common names of birds (e.g., the Allen's hummingbird), it turns out they almost always center on a person's last name when honoring a man (in this case, Charles Andrew Allen). If a bird is named after a woman—in this case I am thinking of Anna's hummingbird, which was named after Anne d'Essling, wife of the ornithologist François Victor Massena, Duke of Rivoli—then it is almost always going to feature her first name. We thus can have "Anna's hummingbird," but never "Charlie's hummingbird" for the male-named species. Both Allen's and Anna's hummingbirds can be seen side by side outside the very same museum building that Dr. Vendetti works in, but one hummingbird was named using one standard, while the other embodies a different standard.

With animal names, traditional science has come up with so many clunkers you wonder if the people doing the naming were *trying* to be obtuse. In South America there is a bat species whose common name is "tailed tailless bat." I am sorry, but who on the nomenclature committee voted for that with a straight face? I understand what the name is supposed to reveal. This species is part of a group of "tailless" (as in, minimally tailed) bats, except this one has a tail. That would be all right, except half of the other tailless bats have tails as well. There is also a bat called the thumbless bat. And yes—sure enough, it has thumbs.

Maybe nomenclature assemblies should recruit more outsiders. Let's invite some spoken word poets into the magic circle, or else a gaggle of skinned-knee skateboarders. Skating slang gives us the "alley-oop," a trick in which the body goes one way while the skateboard spins the other way. What a delicious word that is, both to say yourself and to hear said aloud. The skating world also gives us ollies, caspers, and five-O grinds. In Civil War slang, a bayonet was an Arkansas toothpick, while noir fiction warns us not to be shivved by the gink's moll, no matter how glam her gams are. She may be a looker (these stories caution us), but inside she's still just a cheap twist. Noir fiction preserved and added to flapper slang from the 1920s, when a flat tire was a dull-witted or disappointing date, and the alderman was a portly man's belly.

After identifying so many gulls and terns based on their small variations in color and wing shape, maybe pelagic birders are the best prepared out of all of us to distinguish these quick-flashing silver flyers. I want the namers of new animals to be more like Mr. Howell, who talks about a category of flying fish he labels the smurfs. These are small—only an inch or two long—and at sea, in flight, they look like silvery bubbles or disks. They are juveniles (we *think*) and they are so featherweight that when trying to fly, they "seemingly are at the mercy of the wind," as Howell explains. In some of the smurfs, the yellow-and-brown tortoiseshell pattern in their wings may help them hide in sargasso weed or other mats of epipelagic vegetation. Whatever the purpose, to me the color pattern makes them look like mermaid corsages.

It's a strange, strange world out there. Not only are the fish catching air, but flying squids exist too. Like flying fish, these squids too trade water for air, except they jet out of the water facing backward. Their hind fins create a front wedge, and tentacle arms interlock to create a large, triangular rear wing. A pulse of high-pressure water propels them, and they glide up to ninety feet before splashing back down. Peter Ryan: "On a calm day you can hear a group 'spiss' as they emerge from the water, each trailing a jet of water."

Counting on my fingers by units of ten thousand, by my estimate we are swinging up close to two million words in English. That total includes all of Shakespeare and Milton and every page of a Boeing repair manual, the babytalk blabber I share with the dogs when I am home alone, and even makes a generous sweep of all the slang in all the middle schools in this fair and varied land.

It's a nice start, this first batch of two million, but when I look around and see things like jet-propelled squids or double hyena flying fish, I do have to wonder if we don't have another four or five million words still to come. This new language will exist not only to name each kind of animal and plant, but to capture the chest-swelling excitement that seeing these animals brings out in us. When you first light a Fourth of July sparkler, it sizzles up into flinchy brightness with a burst of white-hot magnesium—nature gives me that feeling too, that rush of bright joy, only I don't have a single good word to describe it. I feel it but can't name it. We all want to fly, and suddenly here is a fish—that least suitable of participants—doing exactly that. *Woo-eeee.* You can't initiate seeing a flying fish; there is no bait to spread in the boat's wake or elk bugle to play on your smartphone. You just have to go out and hope that you're looking in the right place when the miracle happens. And when it does happen, it feels like—well, *something*, anyway. Phrases like "wowza" or "that was *so* cool" are third-rate approximations of the expression I am groping for right now. I don't want to curse— "hot damn!" is dated anyway—nor do I want to sound like a Tudor courtier—"a firksome great fish, huzzah!"—so I am waiting for an option that will help me express surprise and rhapsody and *did you guys see that?* all in a single one- or two-syllable bundle.

We can do it, I am sure. Post hoc therefore, but come on, if we can invent 148 colors of Crayola (plus glitter, pearl, and metallic), then we can come up with good names for the times when our hearts play the sousaphone, and our smiles are so big they could star in their own ads for toothpaste. I am not offering to show the way—my own language skills are iffy at best—but I am offering to be part of the focus group that goes out and has the cool experiences in the first place, and then trials the new words to see which ones fit best. Assuming that there are sixty or eighty or a hundred kinds of flying fish in the world, let's start with seeing those.

Do you have a boat fueled up and ready to head out?

Message me on my website. I will meet you at the dock. Cowabunga, my loyal dudes, mucho mucho cowabunga.

HOW TO BE AN ALBATROSS

Lesson 1, *How to Be an Albatross*

1. A pox on Samuel Taylor Coleridge, whose *Rime of the Ancient Mariner* is the single-most tedious poem in the long, tedious history of British literature.

2. "Few things are more inspiring, or more humbling, than to be far out at sea in rough weather, hanging onto the railing as an albatross glides effortlessly by." (Cornell's *Birds of the World* website)

3. Sleeping: Done while flying, *maybe*, with one half of the brain switched off at a time. Or maybe not, and if not, done the regular way, dozing on the water, especially on moonless nights when it's too dark to hunt. Or maybe they do it both ways, or maybe some third way nobody has witnessed yet. Albatrosses do not spend a lot of time in MRI machines or sleep labs; we don't really know how their cognition is wired.

4. Sex: Yes, but only after a long apprenticeship at sea. After three or five or ten years, once an albatross comes back to the natal island, it may find itself braying, whistling, or grunting, but even if it is the first time, it knows what to do.

5. The math behind flight is straightforward. Flight is a bird's weight in Newtons (mass times gravity) supported by the total area of its wings. All albatrosses have medium-to-low wing loading, which means they don't weigh much but have long wings. Frigatebirds and sooty terns have even lower wing loading, as we will review in a moment. All three though are wind-riding champions. The sooty tern is black and white and when not breeding stays at sea, flying almost constantly. It sometimes lands on floating debris or the backs of sleeping sea turtles. Frigatebirds are larger, blacker, more cross shaped. Like the albatross, it has large surface area and low mass. For albatrosses, technical specs come out to aspect ratios of thirteen, fourteen, or fifteen to one, meaning that for every inch of width, there will be fifteen inches of length.

6. In other words, an albatross is built like a sailplane.

7. Or like the blades of a wind turbine.

8. Frigatebirds, terns, albatrosses—here is one way to do it. The goal is to glide at a shallow angle to maximize forward movement with minimum drop. At its most basic: upward lift versus downward force of gravity. Start by flying into headwinds to increase altitude, followed by a downward path moving with the wind to gain momentum. Rise and repeat.

9. All of this is possible because the velocity of winds blowing across the sea surface increases with the height off the water due to something called the boundary layer effect. Small birds have to flap to fly. If you watch them closely, it's more like flap-glide-drop, flap-glide-drop, little bursts of effort with the body pulled in like a bullet during the glide. But that does not last long, and they need to flap every other second or it becomes not flying but stalling, dropping, crashing. To be an albatross, flying is all about small adjustments in body and wing orientation, little micro tensions that change angles every minute, even every second. If you are an albatross, you are not flapping (not often, anyway); you are cutting and wheeling, rising and falling, working the wind like a master smith honing a sword's edge.

10. Albatross ergonomics: Lift with your back, not your knees. To lock wings into place, it is all about that *click* as wing tendons slot into perfect tension. How easy it all is can be measured in terms of a unit of energy, like how much effort it takes for an albatross to sit still, breathing on a nest and maybe guarding its single egg. If being at rest is "state X," flying usually requires less than 2X effort, or maybe 3X if trying to take off and there's no headwind to provide lift. That is rare, though. Flying should never feel like hard work; if it does, and if you're a seabird, you're doing it wrong.

11. And thus an albatross annihilates distance. In fact, there is no longer such a thing as distance: If there are squid at point B, and if you are an albatross and you are at point A, then go over to B. They ask each other, "Is it a thousand miles each way? So what. Have a nice trip. See you when you get back."

12. A warbler catches a worm or moth, eats it, flies back to the nest, pukes it into the baby's eager gape. Elapsed time, capture to transfer, is maybe five minutes. How can you do that chick barf feeding given that a petrel or albatross may be gone from the nest for days and days, hunting across oceanic distances? All the food would be digested by the time parent bird is back at the jobsite. Solution: turn food into a concentrated oil, then regurgitate that. Penguins do it a different way. They too feed far away, but they shut down digestion completely and use the stomach as a holding bag. They bring back the food in its original format, weeks later. What about boobies? Not studied, but they forage closer to the nest like gulls, so probably come back with whole fish or semi-digested fish mush. The albatross method, to concentrate food as oil, has another advantage, since it arms the chick with something to spew when fending off foxes and skuas. It doesn't always work, but better a limited defense than none at all.

13. Question: Who would win, condor versus albatross? Answer: albatross.

14. That is because (a) the albatrosses have the widest wingspans of any bird, wider even than Andean condor; (b) they don't

have to eat putrid sheep; and (c) multiplicity. There are only two kinds of condor in the world but twenty-four species of albatross. Obviously the wind gods prefer the mollymawks over the sheep-eaters.

15. As of 2024, Wisdom, a female Laysan albatross nesting on Midway Atoll, leg band Z333 (white numbers in red plastic), is at least seventy-three years old, making her the world's oldest-known living bird. To quote a staff biologist, "There is no reason to think she is truly the oldest one. She is just the oldest one *we happen to know about*."

16. Sabbatical: What biologists call a nonbreeding year. Court, mate, raise a chick, and then skip the following nesting season, taking a sabbatical year to feed, rest, and explore the world. A year of parenting followed by a year of recovery. Many of us would agree that sounds very sensible.

17. There is a species of albatross called the wandering albatross— no, sorry. Start over. There once was a species of albatross called for a long time the wandering albatross, which is what I still call it inside my mind, but as a species it has turned out to be part of a cohort of four related species: the snowy albatross, the Tristan albatross, the Antipodean albatross, and the Amsterdam albatross. ("Amsterdam" in this case means an island in the Indian Ocean, not the European city.) Of these four recently parsed species, the former wandering one is now the snowy one. And snowy albatrosses, like many seabirds, have brown irises. Yet in a genetic quirk, about 1 percent end up with blue eyes. Why does it happen, and does that feature make them more (less?) attractive as mates? Nobody knows.

18. My father had rare eyes, statistically, since his were translucent gray—gray with a slight rinse of blue. I want to think that is what my mother noticed first about him. More likely it was that his suit didn't fit, or that he had cut himself shaving. They had met at church, two late bloomers, stiff and awkward. Perhaps he shook her hand; neither was a hugging kind of person. When he met her, he had most recently been working in a tire factory, and before that he had been a cowboy, and before that he had been in the war, what he always called the service. After they were married, he kept a book from the navy days that was an overview of a sailor's duties, from tying knots to cleaning a gun to how to say things with signal flags. Too young to read, I was fascinated by the pictures. I too wanted to say things with flags the size of my body, sharing my thoughts and requests with faraway men trapped on other islands, trapped in other families. One day, though, when I went to look at the flags, the book was gone. My father had thrown it out. The story he told me halfway made sense. He said that the author of the book had died of tuberculosis, and he told me that the book was dirty; it would give me a disease. I was sad to see it go, but I understood. I knew about TB because a neighbor had caught the same thing. I hadn't known that TB came from books, but I was glad our house was safe. As an adult, I realized there

was indeed something contaminated with that book, but it had nothing to do with bacterial infections of the lungs. Either specific pages or maybe the entire book made my father anxious, and he could not have it in the house with us anymore. Strange man, my father, and why did he even bring up the tuberculosis story at all? I think he really did believe it, at least at some level. I keep replaying the old scenes in my mind, finding new clues with each visit.

19. In Yorkshire on the northeast coast of Britain, a black-browed albatross was visible for several years, roosting in a gannet colony. This is an albatross native only to the southern hemisphere, so how it got there, nobody knows. There's a void around the equator, the doldrums, where the winds stop; that invisible barrier is what keeps southern hemisphere species from showing up in the north. But whether there was a freak storm or a helpful boat, somehow this one stray did make it into the North Atlantic. What it was doing once it found itself off the coast of Britain was looking for sex. Gannets were the most albatrossy things it could find, so it went to their colony. There was an Australian gannet that got lost in the southwest Indian Ocean once and hung out with king penguins. We know that with humans, *Homo neanderthalensis* and *Homo sapiens* comingled so often that elder DNA lives on inside of us. The only thing stronger than the urge to get away from your kids for an hour, a day, a week, is the urge to hook up with somebody (no matter how unsuitable) and make more kids.

20. An albatross can dive underwater. In fact, the word "albatross" (by way of Spanish by way of Arabic) comes from a root word for diving. It can go fifty feet down, if that is where the fish are. Mostly surface feeding though, that is their preferred approach, and if, say, there is a dead whale, then sharks and albatrosses will stay with the carcass for days, even weeks, swimming beside it as the dead whale floats on the surface, everybody pulling off chunks, with more and more other birds joining in, each new-comer following the oil slick upwind to its greasy, swollen source.

21. Thus, to see an albatross, find a dead whale. If that isn't possible, follow the fishing boats.

22. Albatrosses mate for life—sort of. It's hard work to court a mate, bond with them, raise that voracious chick. So yes, they pair up, on average, for life, or at least until one of them dies, but at the same time, tests show that between 10 to 20 percent of chicks are born to a genetic father who is not the female's social partner. Like humans, albatrosses stray. And like humans, albatrosses get divorced. This can be amicable—both partners stay in the same colony, but find new mates directly—or more acrimonious, when one of the pair is with a new partner right away and the other is alone, that year and sometimes many years successively. And other times they divorce only temporarily, and after a season with other mates, both come back to their original partnership once again. What triggers these changes and how it all is negotiated, nobody knows.

23. If albatrosses are among the best birds of the world for having the most lift for least weight, one seabird beats even the albatrosses. That is the frigatebird. In flight, shaped like a crucifix, or to be exact, shaped like a crucifix with a forked tail. They float there, sometimes so high up they look smaller than the dot of an *i* made with a sharp pencil. Males are black, with a slit-throat red cravat. Melville called them sky hawks. Here's the frigatebird's party trick: the total weight of their skeleton is less than the total weight of all their feathers. If you've ever held a balsa wood model plane, or heck, even a regular paper airplane made out of notebook sheets, you have an idea of how minimal their skeleton really is. It is the size of a full-grown person, but it does not weigh any more than a wet cat. In fact, the frigatebird is such a blue-ribbon flying champion, it can stay aloft for months. To do so, it has to sleep on the wing. Besides the half-brain-at-a-time on-off switch, a frigatebird also takes micro naps of twenty to sixty seconds, when both sides of the brain shut down at the same time.

24. All that is to say, the albatrosses and frigatebirds are the kinds of birds that have their bodies so finely tuned, so weight efficient and distance achieving, that in comparison, humans are just dull lumps of dense clay.

25. In that case, maybe we should aspire not to be the wind bird, but the wind.

Lesson 2, *How to Be the Wind*

25. Tawhirimatea took to wife Paraweranui and begat a host of Wind Children. Collectively they are known as the Aputahi-rangi. Offshore, Tini o Matangi-nui and Tini o Mataruwai are the names for all the winds of Mahora-nui-atea, the vast ocean spaces.

24. In another story, Maui succeeded in confining the winds in a cave, the mouth of which was blocked with stones to prevent them from escaping. The west wind was the only one that eluded all Maui's efforts to catch it, and so many times Maui the Wind-Seeker rode forth on the north and south winds in search of the mauru that abides in the realm of Tahu-makaka-nui. At certain times the blockading stones were removed from the cave, and the Wind Children were allowed to come forth and roam the great sea plains of Mahora-nui-atea, to frolic on the rolling plaza of Hinemoana, the Ocean Maid, and to chase the Whanau kapua, the Cloud Children—chase them and drive them away far beyond the hanging sky.

23. We know the clouds have broken up and blown away. Do you have a better explanation?

22. Moody kind of weather, the wind. Unreliable. From calm to medium to gale force and back again, sometimes in twenty-four or thirty-six hours. Strongest-known surface wind, in miles per hour: 253. Probably stronger winds than even that inside a tornado, but the instruments never survive.

21. There is never a time when the wind is not blowing somewhere. The sea is never fully calm. Never level.

20. And on Saturn, thousand-mile-per-hour winds, and on Jupiter, green lightning: every time it seems like we're the best, somebody else is better.

19. Frigatebirds come to the desert after storms. Not often, but it happens; I have seen them. And there is a record of an albatross in Palm Springs once, heading north into the Beaumont Pass above Interstate 10. Presumably it had been blown from the Sea of Cortez into the Salton Sea, and from there was trying to find a way back out to the Pacific. Things to watch for, if making the drive from Los Angeles out to the high desert—count the cannabis billboards, pity the has-been bands headlining casinos, note the two concrete dinosaurs by the outlet mall, and above all, keep an eye peeled for northbound albatrosses.

18. Another thing to watch for: burning things. I once saw a car on fire on Interstate 10 right at Beaumont Pass, parked on the shoulder, fully engulfed. Black, late-model SUV. Orange flames from a black car against a brown desert, with a concrete T-Rex visible in the background. Who needs to go to the movies when real life is so vivid? That's one question I ask myself almost every day of the year. In this case I did not stop—nobody else did either—but I did call it in, since one of my lesser skills is that I do know how to give good directions. I was going seventy, maybe seventy-five, so in the sixty seconds that the 911 transaction took, I was already one mile past the scene, and soon it was two. Then three miles. Four. A sixty-thousand-dollar car was burning down to the axles, and it had gone by so quickly it had barely made an impression. I suppose my life, my problems, are like that to everybody else.

17. Beaumont Pass was and is a major wind turbine site. Tom Cruise filmed an initial scene of *Rain Man* there; it serves as the abandoned family homestead in the Bradley Cooper version of *A Star Is Born*. Call it the poster site for clean energy. Assertion: all energy is dirty energy. Our leisure always costs—the only question is who will pay. Wind turbines in that perpetually windy corridor help charge up your neighbor's new car but at the same time are spinning wheels of death if you happen to be, say, a golden eagle or a migrating bat. The numbers are appalling—in North America, 500,000 bats are killed a year? Twice that? And 150,000 to 400,000 birds? (Or more, way more?) At night the hazard lights blinking from the turbine tops blend to form a malevolent fortress of red eyes, eyes that are always awake, always watching.

16. Yet wind turbines at sea may be all right. Or less bad than the ones on land. The pillars of the sea turbine towers will provide

substrate for kelp and anemones and so on, and if you build them tall enough and light them wisely enough, at-sea turbines may actually be far enough above the surface of the water so the shearwaters and gulls will go under the blades, not through. That is the theory, anyway. Yet some studies say that ocean windfarms are as bad as the land ones.

15. But if so, oh well, since we have to get our juice from somewhere, right?

14. Another idea, and this is from recent work in Norway: if you paint one turbine blade black—not all three blades white but two white, one black—it creates a better "smear," visually. All of a sudden, a bird can see that there is something there, lift up or go around. For bats, one possible deterrence is to jam their radar. They need to echolocate to navigate; if there is so much static they can't get close to a turbine, the bat will back off, swing around, stay in airspace that it can perceive cleanly enough to navigate through. "Mitigation," a word energy companies claim tastes like vinegar. A troublesome word, but necessary. All energy is dirty energy, but some energy is dirtier than other energy.

13. Clean and bright plane. Such long wings. I went up in an all-white sailplane once, using a service out of the Lone Pine Airport, Eastern Sierra. Late fall, crystalline day: the tow plane took us up to eye level with Mount Whitney and cut us loose. Nothing but sky and silence, view and awe. There was so much lift

we must have been up for hours, making slow banking turns as the basin and range of the Great Basin stretched to infinity on one side, and the most beautiful mountains in the world glowed white and black under the season's first snow on the other. To be in the middle of the air but to have no sound at all, no vibration, to be flying based on wingspan and fiberglass, was to be as close to a bird's ease as humans will ever get.

12. I have never gone up again. I know it cannot possibly ever be a better day, and I do so much in life that is always second best, always compromised or rushed or dirty from this or that faulty connection, I have promised to keep aside a few experiences that start and end with perfection.

11. A friend had a hang glider, a tandem model, and was going to give me lessons. He crashed and died not long after we made our plans, and in fact died flying the exact same wing we were going to use. I went to his funeral; his friends all showed up in trucks with their own gliders tied to roof racks, since most had been out doing their favorite sport the morning before the service. For me, I figured somebody Up There was giving me a sign, and I did not follow up with lessons from somebody else. While that probably was the smart choice, I still feel regret about making it.

10. It is easy to want to die the hero's death; we celebrate mountain climbers and smokejumpers and motorcycle Hall of Famers, and we like it in movies when the good guys chase the bad guys across

the roofs of moving trains and catch the escaping airplanes by hanging on to the edge of the wing. Some small, dim part of us knows, "It's called 'the laws of physics'—that stunt is pure bullshit." And another, larger part of us is glad that the hero has tried, and even more glad when they make it. *The Fast and the Furious*—that is the soundtrack to my life, even if I can never say so out loud. I often talk about driving my beater 4x4, but I also have my town car, a zippy little white job, and that car's speedometer promises it can hit 180. I have never tried to find out if it can or not, but I will be honest—most days I want to.

9.   Longest verified circuit in a hang glider is 222 miles. That is not longest flight in a straight line (which is nearly 500 miles), but in terms of being able to gain altitude, do a traverse, work your way over and through and around mountains, and come back to the same starting point, then that's the record—222 miles in one day. That journey was done in the Owens Valley, along the Sierra crest, in eastern California. A seabird might hear about that and say, "World record? Good start, junior. Come back when it is ten times that distance, fifty times, one hundred times."

8.   Frigatebirds rise on thermals up to a few thousand feet, then glide back toward the surface, then do it again: steep climbs and long glides. If they go up as high as fifteen thousand feet (riding the updrafts of cumulus clouds), they can use that launching point to glide for fifty miles before needing to kettle back up. Usually they rise less high and glide less far, but it's

still effortless. Rise, fall, rise: during a given day, a frigatebird uses almost no energy at all.

7. Albatrosses, frigatebirds, and in fact almost all birds breathe differently than we do. Air circulates for several intake-outtake breath cycles, filling not just lungs but also pre- and postlung air sacs, until every smidge of breathable oxygen has been extracted. Birds live in air in ways a mammal never can, and air lives in them, filling them and supporting them. When the air finally leaves them, it does so fully depleted, fully spent.

6. It seems almost unfair, how free they all are. Even on the nesting grounds, if an albatross suddenly thinks, "To heck with this nonsense," it only has to spread its wings, turn into the wind, and be lifted away.

5. But of course, we can do that too; we only have to give ourselves permission to try.

4. There was a person I thought I was in love with, whom I thought I was linked to forever. We had been married in a church, in front of God and in front of my family and my friends—those friends she had decided I was allowed to have—and inside the boundaries of my faith, that meant the marriage was permanent, eternal. We had no money, and for the wedding decorations had ended up using leftover flowers from a funeral. Her dress had cost thirty-five dollars; my mother had paid for it, and

also bought us six cases of Henry Weinhard's for the reception. For ten years this person and I were inextricable, joined as tightly as foot and shoe, as fist and slap, as blade and sheath. Even after she slept with one of her students, even after she smashed a bottle and tried to stab me, even after her illicit boyfriend crashed my car, even after she held a gun to my head, still I stayed with her. I had made a promise before God—that was my idea of marriage. So strange now, to think about how small my world was at the end, how impossible it was to look past the circle of black chalk she had drawn around me, commanding me to stand in the center and never step out. A single circle, that was my radius, and it made sense at the time, even as I knew, or suspected that I knew, that there were other molecules of air besides the few that she allocated to me to breathe and eat each day. I knew but could not act.

3. Perhaps you have seen pictures of somebody on a cliff edge— there used to be a Patagonia ad like this—with the person leaning forward, arms out, coat billowing, leaning over the cliff and looking into space and being held in place by the wind.

2. I used to wonder how they could do that—what did they know that birds know and most of us do not know, or if we know, we have trouble acting on? And the answer is (to quote Edward Tufte quoting Joanne Cheung),

1. "The horizon is not a line / it is a space."

THE LAZARUS BIRDS OF NORFOLK ISLAND

Islands make good prisons. Think of San Francisco's Alcatraz, or Devil's Island in French Guiana, or the famous case of defeated Napoleon, who died in exile on Saint Helena in the South Atlantic. The first island he had been imprisoned on, Elba, wasn't good enough—it was too close to France, and he escaped from that one. The British had to put him someplace they could guarantee he would stay put, so they stuck him on a volcanic dot on the map a thousand miles west of Angola.

Another off-the-edge-of-the-map British prison was on Norfolk Island, isolated in the Pacific eight hundred miles from the coast of Australia. Before the British claimed it, there had been Pacific Islanders living there, but they had come, settled, and departed, leaving behind little by way of monuments or cultivation. They may have been Polynesians, the master mariners who also settled New Zealand and Hawaii and Easter Island, or perhaps Melanesians from New Caledonia. By the 1500s, they all seem to have left, and the island had reverted to its wild state: green and flat and hoisted out of the sea by steep cliffs.

London at the end of the eighteenth century was a city with too much crime and too few jobs, and so to the authorities, Norfolk Island seemed like an ideal solution. Once you dumped the miscreants there, the only way back off the island was by ship, and nobody in the Admiralty was planning on being profligate with those. Robert Hughes in *The Fatal Shore*, his history of Australia, explains it this way: "Magnificent in scenery, Norfolk Island was also a natural prison, harborless, cliff-bound, and girdled with reefs on which the long Pacific swells broke with a ragged, monotonous booming."

The problem with Norfolk as a penal colony is that there is no food, or rather there is no food for people who do not have boats for fishing and a knowledge of the local botany. If you want to grow European crops like apples or potatoes or wheat, this is very much the wrong soil, the wrong climate. In any case, an apple orchard takes years to nurture and bring to fruit; an overnight crop it is not.

Despite these immutable realities, starting in 1788, successive groups of marginalized, quasi-British citizens—prisoners and soldiers alike—were sent to Norfolk to serve out their respective terms of enlistment. Cut off from everything, including the newly established penal colony of Australia, all parties felt like they had been banished past the edge of hope and memory.

How many people were there I can't quite make out; I am guessing several hundred. Did the guards have muskets? One presumes yes, and some cutlasses too, and maybe even uniforms. It did not matter; everybody was trapped on the same foodless island. You could kidnap a guard and nobody would care, since everybody was equally imprisoned. The first crops failed, burned by salt spray and eaten by rats, parrots, and cutworms. The only supply ship that was coming smashed on the reef, and all its cargo was lost. The ship's crew was saved, but that meant the castaways joined an already gaunt population. No other ships were going to come.

In exiling these wretches past the supply chain of a still-nascent empire, London was enacting a universal if slow-acting death sentence. What happened to the prisoners once they were deported to the Southern Hemisphere was not of any particular concern to the authorities. The prisoners were expected to fend for themselves, even though they had no training and no equipment. Most were guilty of midlevel crimes like theft, and most came from urban slums; the colonists did not include crofters or journeymen carpenters. Skilled tradesmen did not steal bread and get banished to Australia.

By 1790, the reality that they would starve to death faced every person on the island. They needed a miracle. And then one night in April, they got one. During the night, as if released by guardian angels, flocks of savior birds began to arrive. These birds, clearly sent by divine intervention, were named by the colonists "providence petrels." The name still is used today.

At sea, this gadfly petrel is a fast, acrobatic flyer, gull sized and steely gray, with double chevrons at the base of the wingtips and a smudgy white muzzle like it has taken first place in a hands-free cake eating contest. Their wings look like a cross between a boomerang and a sword, sharp and flexible. They are aerial masters and lift anybody's heart when you see them wheeling and arcing through the ocean air.

On land, having grown accustomed to thousands of years of freedom from mammalian threats, the petrel is calm and trusting and easily bashed to death. This is true for most seabirds in the Pacific. Before people arrived, on each island or atoll or remote cliff, there were no rats, no feral house cats, no mice, no goats. Species like shearwaters and tropicbirds may spend most of their lives at sea, but sooner or later they need to lay eggs and raise chicks. They need—however temporarily—refuges on land. They need places like Norfolk Island.

The providence petrels came back from sea to mate and raise young the same as they had every year, except this time there was a colony of desperately hungry people waiting for them.

The British were merciless. We do not have accurate logbooks—there were no resident historians, no park rangers, no field biologists, no country parsons. As the best guess it seems that up to five thousand petrels were killed each night, and we know that the colonists took not just adults, but eggs and chicks too. Most ground-nesting seabirds bite and lunge, or in the case of cliff-nesting fulmars, at least try to projectile vomit into the face of an intruder. Providence petrels did none of those things. Some people were so covetous of the eggs they cut open living birds looking for them, and if there were no eggs, threw the birds down to bleed to death. This butchery continued each breeding season for the first three years and lasted until the settlement started to become self-sufficient in raising their own crops and livestock.

Lucky thing for them that this was so, since by then the bird population was all but wiped out. A few years later, the last surviving petrels were still trying to nest, but cats, rats, mice, goats, and pigs soon completed the birds' extirpation.

Over a million birds had been killed.

Yet somehow a splinter population was able to nest on Lord Howe Island, five hundred miles south of Norfolk. How the birds found it and if an outpost colony had always been there, that part of the story remains undocumented. All we know is that Lord Howe became an accidental refuge, despite eventual arrival there too of European pigs, cats, goats, mice, and rats, and despite the Australian masked owl, which had been intentionally released to control rats and instead hunted providence petrels. There is no bad situation that humans cannot make worse.

On Norfolk Island, all that struggle, all that misery, served no purpose in the end. The penal colony was abandoned. It was too difficult to maintain, even for an empire as large and cruel as Britain's. That first absence lasted twenty years, then the prison was reestablished, and then it was abandoned again, the last time in the 1850s.

This is where I come into the story, or I do after we start over, this time on Pitcairn Island, four thousand miles away.

2.

In 1790, at the same time that the Norfolk Island prisoners were on the brink of disaster, eleven Tahitian women, six Tahitian men, nine Anglo-European sailors, and one Tahitian baby tried to start a new community on Pitcairn Island in an equally remote part of the Pacific Ocean. The Polynesians—the women, anyway—had been enslaved by the Europeans, all of whom were fugitives, since they were the surviving mutineers from the HMS *Bounty*. If caught, they would be hanged. To hide the evidence, the mutineers burned their ship.

The *Bounty* mutineers were sailors on the *Bounty* in the first place because the ship had been sent to Tahiti to acquire breadfruit saplings, since the British Empire was looking for cheap food to feed enslaved people on Caribbean sugarcane plantations. You might ask why they didn't grow local crops instead. This is a reasonable enough proposition, but as with Norfolk, the islands of the Caribbean were good at sustaining low-density master foragers who know how to fish and live lightly on the land. Precontact, the Caribs and other indigenous tribes thrived. Once Europeans enslaved them, that knowledge base was lost. In any case, once you start to build a slave-based economy, you have a lot of people concentrated in one place, and they need to be fed quickly and easily, and fed with crops that will not siphon off too much of the labor force into cultivation. The British authorities decided they needed to plant breadfruit trees, and to do that, they needed to bring them from the Pacific to the Caribbean.

As a plant, the breadfruit tree is tall, spindly, and vaguely tropical; the fruit looks like a distended green grapefruit with the same pebbly, dinosaur-skin rind you see on the durian or jackfruit. Broadly speaking (depending how you prepare it), breadfruit tastes like potato. When the mutiny happened, the breadfruit seedlings that Bligh and crew had carefully nurtured were already on board for the trip home. Once the rebels took the *Bounty*, they threw all one thousand saplings overboard, intending to repudiate every single aspect of their original mission.

In popular culture today, the name Captain Bligh stands in as shorthand for a tyrant, or at best an imperious martinet. Revisionist historians exonerate him, saying, "He could be a bit of a knob, sure, but given maritime culture, he wasn't so bad." At the time of the actual event, the *Bounty*'s forty-person crew was evenly split; half mutinied and half wanted to stay loyal. The mutineers won.

Bligh and those loyal to him were set adrift. In a brilliant act of bold, open water navigation, Bligh and exiles made it three thousand miles across the Pacific in a small boat, ending up in Timor, in present-day Indonesia. It took forty-seven days. Once there, in what was then called the Dutch East Indies, some of the crew died, some dispersed, and some made it back to England. After a hearing in London, Bligh was assigned a new ship and was sent back to the Pacific for more breadfruit trees, since feeding enslaved people cheaply remained an unmet goal. That second voyage was successful, and breadfruit now occurs worldwide.

I first ate breadfruit when I flew to Grand Cayman Island on a last-minute, heavily discounted ticket. It costs money to go to tropical islands, and this hasty trip was being done on a budget smaller than the aglet on the tip of a shoestring. I could not afford a hotel and had vague plans about sleeping on the beach, but I was lucky enough to fall in with a couple who let me crash on their sofa for five dollars a night, meals included. "He Hath Founded It upon the Seas"—a line from Psalms 24 is on the Cayman coat of arms, and I still have the cloth patch sewn to my old backpack. It's hanging in the rafters of my garage, near where I keep my maps.

The man of the house, Noble, was an Afro-Cuban dockworker

with Burt Reynolds sideburns and a Doc Holliday stinger. Barrel-chested and proud-paunched, he did not believe in wearing shirts. "A man's body has got to BREATHE," he told me, thumping his hirsute man-breasts. I got the idea that even wearing swim trunks around the house was a concession he had only reluctantly agreed to make due to my unplanned arrival.

Wife Margot worked three days a week in a bookstore and was in charge of meals. (I helped by washing up.) The first night Margot cooked wearing a blue floral house dress with a hibiscus flower decorating her henna-streaked hair. Dinner was conch fritters and breadfruit salad, and both foods were new to me. There was spiced mango for dessert. Theirs was a basic menu. If they couldn't grow it or catch it, they didn't serve it. Store food was for rich people.

As we got ready for supper, Noble decided to show off. He got out his .22 rifle, and as Margot shook her head ("ammunition is expensive"), he stepped into the yard, took a steadying breath, and with three quick *pops*, he shot a breadfruit right out of the tree. I think about that whenever I drive past "U-Pick" apple orchards— "Excuse me, is this a bring your own gun kind of place, or do you have loaners?"

Any time I eat breadfruit now, all these weary years later, tasting it reminds me that Noble and Margot were very kind to me, and when I look back at pictures of myself from back then, I am so childish and virginal and eager to quote the Bible that I look like a somewhat addled golden retriever. I had long hair and brown legs and the world, I thought, was going to be forever easy, forever interesting.

3.

These days Norfolk Island is an Australian state called an "external territory"; nonresidents need a visa to land. There are some slow-burning independence movements and vague interest in seceding from Australia and joining New Zealand. Nobody plans to go to war over it any time soon. It has such a small population that when I visited, I saw a car with number plate 287—that was it, three digits only, black numbers on yellow metal. Wave to your neighbor as you pass on the highway.

Norfolk Island is also now the home of the *Bounty* mutineers' descendants. With their resources running out, in 1856 they moved from Pitcairn to the recently abandoned penal colony on Norfolk. They sailed there on a British-supplied clipper, the HMS *Morayshire*, newly built out of high-grade teak at a shipyard in Burma. Most of the Pitcairn descendants still live there.

Besides exploring history and besides encountering the namesake Norfolk Island pine in its native habitat, most birders want to see the endemic robin and the flashy red parrots and a startlingly green pigeon called emerald dove. I visited Norfolk as part of a seabird study trip and hope to go back and to stay longer. The air was humid and tropical and yet cool, like how it feels inside a walk-in fridge at a high-end florist's shop. At lunch a vivacious and generous resident, Margaret, tried to teach me Norfolk, the native creole. As the people in other isolated communities have done, Norfolkians have developed their own language. With birds, for example, wedge-tailed shearwaters are called ghost birds, while noddies are titeracks, and sooty terns are called whale birds. All rails (no matter the species) are tarlers. Norfolk combines archaic English—a little songbird known as a white-eye in the field guide is locally called a grinnell—with Pacific Island words, such as *nuffka*, the sacred kingfisher.

I love languages but have no aptitude for learning them. Margaret and I did not get much past *wataweih* ("hello" or "g'day"), plus a sentence about fossicking for shellfish at low tide that I didn't write down in my journal and now can't remember.

Margaret's last name was Christian, as in Fletcher Christian, as in one of the original Pitcairn mutineers. Mel Gibson played him in 1984's *The Bounty*. Through Fletcher Christian, Margaret is an eighth-generation *Bounty* descendant. She still has oral traditions received from her matriarchal heritage, along with ambiguous, multiracial features. She told me she is assumed to be Polynesian when in Tahiti, Maori in New Zealand, and Aboriginal in Australia. She is all of these and none of these; as she explains it, her heritage is Norfolk Islander by way of Pitcairn, but mostly she is just a race called herself, as are we all.

During the initial slaughter of providence petrels in the 1790s, convicts and wardens on Norfolk were too busy killing birds to set aside even a single carcass to be preserved for scientific study. Because of that, this animal was not published as a formal species until fifty years later. A London naturalist named John Gould wrote up the species description, calling it Solander's petrel. His scientific name survives today, *Pterodroma solandri*, remembering Daniel Solander, a Swedish biologist who traveled with Captain Cook on the *Endeavor*. Gould honored Solander in error, since this bird was not collected by Captain Cook or any of his crew; their colonial excursions into the Pacific predated the Norfolk Island colony by thirty years. Gould did not know the real story about the bird or its near demise. That came out later.

4.

We cannot reverse imperialism, but we can try to disentangle its many strands. (David Antin: "All the stupid history of human culture is embedded in our language—along with most of its brilliance.") Ornithology is now entering a yearslong project of rethinking why we name things the way that we do. In the Pacific, trying to decolonize language means trying to honor a natural history that was tainted by capital *H* History in indelible ways.

Even trying to write a neutral sentence seems impossible. One verb used to describe Pacific history is blackbirding—it means to kidnap Pacific Islanders to sell elsewhere as enslaved labor. I debated using it to reference the Polynesian women taken to Pitcairn with the *Bounty* mutineers. In the end, I didn't—and I also didn't use "consort" to describe them, which is fussy and inaccurate, yet is a word that still turns up in historical narratives. There can be Albert, who was Queen Victoria's prince consort; there can be the idea of not consorting with riffraff; and there is a transactional version, in which a consort is like an escort or courtesan, which is to say, a high-end sex worker. None of those connotations are quite right for what happened on Pitcairn. Reconstructing culture and nature after the fact feels like trying to mend a silk wedding dress after it has been cut into strips and used as shop rags by a particularly messy pit crew at a discount Quiki-Lube. Now that the shop has gone out of business, nothing is as it once was, and there are oil stains everywhere.

In the case of the providence petrel, it did survive, thanks to accidental sanctuaries such as Lord Howe Island. There are about thirty thousand pairs today. If we want to find a better name though, the original name given by Gould, "Solander's petrel," doesn't work, since Solander was a botanist who had nothing to do with this bird. Keeping the current name of "providence" commemorates people who tried to exterminate the birds, not protect them. Nor is there an extant, precolonial local name, since all the Pacific Islanders' pre-contact names are now lost. Using the place as a bird name does not work either. Modern-day Norfolk and Lord Howe Islands were both named after British sponsors who had no interest in (or even awareness of) these antipodal colonies.

Plumage details provide little to go by; one field guide repeats the color "brownish gray" a lot. Nor can we name the petrel after behavioral traits. It flies with soaring arcs and plunging dips, but then so does every other *Pterodroma* petrel. The range too is very "ish"—Western Central Pacificish, though it wanders as far south as Tasmania and as far north as Japan and the Aleutians. Sometimes winds and whims take it as far east as Hawaii. It is a bird of the open ocean, linked to no specific, discrete place. Nor do vocalizations give it a clear identity. According to Cornell's *Birds of the World*, it has two principal vocalizations: "a rapidly repeated *kik-kik-kik* in flight and higher-pitched, warbling *ker-rer-kuk-kuk-kuk-ker-rer* given from the ground." Neither sound is commonly experienced by humans.

If English were an agglutinate language like Estonian or Inuktitut, we might be able to cook up a polyhyphenate compound, something like "Poorly Known Seabird that Once Was Almost Made Dead due to Human Stupidity but Now Sort Of, Kind Of, Is on Its Way Back." A shorter version might be to call it the "Lazarus bird" in English, based on its final return from the dead. Even if we ignore the cultural problems caused by a New Testament allusion, there are other back-from-the-dead seabird species as well, such as the short-tailed albatross. In that sense, almost all seabirds are Lazarus birds.

It is a truism to say that the birds don't care what we call them. We care though, and we care a lot. The names are for us; they contribute to that swirling nebula of "who we are." We want good words for birds because we want good words, full stop. Hamlet tells us there is providence in the fall of a sparrow. Providence too then in the naming of that sparrow, and in naming the accipiter that wants to catch it, the swift-as-an-arrow sparrowhawk. Bird as images, as three-dimensional objects, live in our eyes, our hearts, and our bird apps, but birds as names—those are even more essential, because those words live inside our mouths, our minds, and our deepest souls. To name something is to own it but also to internalize it, and thus to blend its linguistic DNA with our own.

With flying fish, as we have seen in another chapter, whimsy drives common names, and that's fine with me. All mimsy will be our borogoves, and let us clap with delight as the mome raths outgrabe.

Moving into history as a source of loan words, to sail widdershins was to be contrary to the sun and prevailing swell, against the current, to be the wrong way around. There is a bold defiance in that: it might match tenacity and crosswind maneuvers to name this seabird species the "widdershins petrel."

Another solution might be even closer at hand. Earlier I mentioned the generous naturalist, Margaret Christian, the woman whose speech and cultural heritage embody multiple heritages. Final offer: I want to name the providence petrel after her. If she demurs, let her nominate some other conservationist with blended traditions and a good heart. In food, there is the locavore movement. Let's extend that idea to our naming traditions, too.

5.

Natural history study struggles to know how to talk about history and imperialism. Should I castigate the colonists on Norfolk Island for wiping out the petrel colonies? They showed no compassion, not even any enlightened self-interest. They were greedy as brutes. But then again, if I look at it another way, these were very mistreated, uneducated people I am getting angry at. That means the prisoners were often victims themselves—forced to become criminals by poverty and lack of social mobility, and then sent to rot in a forgotten corner of a vast and indifferent empire. On Norfolk they would have died if providence had not given them this magical gift of free food; the imperial system itself had failed them, and only blind luck (and a lack of ethical restraint) allowed people to survive. The deportation system was so cruel that under some sentences, even once you served your time you could still be executed for returning home to England to see your family—a situation Charles Dickens describes in his novel *Great Expectations.*

Sort of a complicated mess, nature study, but then so is the world itself. One bird that shares the same ocean is the Gould's petrel. It is black and white, and like other *Pterodroma* petrels, it has a bounding, wheeling flight. I never expected to be able to see one (other than in a museum). That is because Gould's petrels live far from land and far from organized pelagic trips, in the vast unknown of the Central Pacific. And it is also because they very nearly became extinct, and even more "nearly extinct" in their near-extinction-ness than the providence petrel. What happened to the Gould's? *Boom, boom, boom*—among other things, one main nesting island, a mile-long green hump called Cabbage Tree Island, was used for target practice in World War II. Further, not-native-to-the-island rabbits, first released in 1906, ate all the bushes the petrels nested under. When the nests were exposed, a kind of Australian crow called the pied currawong ate the petrel chicks. Owls and peregrine falcons ate both fledglings and adults. All in all, Gould's petrel seemed completely doomed. Odds of survival were a thousand to one.

In a final Hail Mary pass, authorities enacted a crazy plan. It included a basic to-do list: cull the rabbits, chase away the butcherbirds, get rid of a sticky, bird-entangling plant called the birdlime tree, and at the same time, translocate some of the petrels to adjacent islets, hoping to build redundancy into the shaky system. If nesting failed on one island, maybe a handful of birds would endure on another island down the coast.

Sometimes longshots work out. Population today, off of Australia and including a subspecies on New Caledonia, now tops two thousand. The species may yet last out the rest of this century.

Saving Gould's petrel in turn helped to save another seabird, the Bermuda petrel or cahow. The word "cahow"—originally spelled *cahowe* and *caohoo*—is a loose transcription of the bird's courtship cries. Same history lesson with the cahow as with so many other seabirds: the population circa 1492 was somewhere around half a million breeding pairs, and yet by the mid-1620s, the species was believed to be extinct. It nested only on Bermuda, nowhere else, and as soon as humans arrived, that was all she wrote. Pigs and rats and the thudding rise and fall of wooden clubs, same as before, same as after. Draw a finger across your throat because it's over.

The cahow went extinct in the 1600s and stayed extinct for three hundred years, all the way until February 22, 1906, when one was accidentally found alive in a crevice on Gurnet Rock, northeastern Bermuda. Found but not found, in that it was misidentified as a different petrel species. That one died in the Bermuda Aquarium, unmated and unknown. Ten years later the big oopsie was figured out, and in 1916 what is now called *Pterodroma cahow* was described to science based on bones in caves, historical accounts, and the 1906 bird's died-in-the-aquarium stuffed body. That specimen is still extant; the American Museum of Natural History in New York keeps it, and if you ask nicely (as I did), you can go see it for yourself.

Other than a few strays that stunned themselves flying into lighthouses and radio masts, no more cahows were discovered for fifty years. An expedition in 1951 found half a dozen nesting pairs on a subislet off the Bermuda mainland. Let bells ring joyfully across plaza and plateau: the cahow was back from the dead.

Except it wasn't.

Take tropicbirds, for instance. Cahows only come ashore to breed. A pair mates for life and uses the same burrow year after year after year, returning to the same few square meters of dirt where they were born. Survival rate for chicks ranges from 25 to 50 percent; once the young fledge, everybody goes back out to sea. The young ones don't come back to land again for three or four years, not until ready to breed, though when they do come back, they unerringly home in on the same splotch of moss and rock they hatched from.

So far, so good: some survived, and in the 1960s, hurricanes had not yet begun to erode their nesting islets. Ah, but cue up the scary music, since here come the tropicbirds. We have met these before; they are white, streamer-tailed, highly buoyant birds—lovely to look at and generally a crowd-pleasing thrill when encountered by bird boats. But they too are burrow nesters, and they're larger than cahows. So, what with nature being red in tooth and claw, the tropicbirds were taking over all the cahow nest sites.

The solution, arrived at by way of trial and error, was (a) to create tropicbird-exclusion baffles, only allowing things the size of a cahow into selected burrows, and (b) to fabricate artificial burrows and place them all over the adjacent hills and cliffs. These have worked well. Artificial concrete tunnel or native marl, it's all the same to a cahow. They are not particular about their geology.

Issue sorted, except then there was still the DDT problem. And the snowy owl problem. And the hurricane problem. And the constant, never-ending, always-there problem of how few cahow were (and still are) left in the world.

A schoolboy now becomes the hero of this story, or rather, somebody who was a schoolboy when he was tagging along on the 1951 expedition but who later became the grown-up David Wingate, OBE, Bermuda's first conservation officer and the savior of the cahow. Mr. Wingate did it all: weighed them, counted them, advocated for them, worried about them, tended them, housed them. He was the one who came up with the correctly designed artificial burrows, and when DDT caused eggshell thinning, he flew to the United States to testify. One year, when an off-course snowy owl arrived on Bermuda and began eating the cahows one by one, Sir David loaded a shotgun and did what had to be done. If the cahow has a single person responsible for getting them from the edge of extinction to "well, who knows, they just might make it," it would be David Wingate.

Even though he felt there was a lot of work still to be done, Wingate was forced into retirement by Bermuda's mandatory age limits. His replacement was his protege, Jeremy Madeiros. One of Wingate's successes had been to revegetate a place called Nonsuch Island. The revegetated island not only protected the endemic skinks and snails, but it was tall enough not to be overwashed by the storm surge of hurricanes. With climate change, hurricanes had been bashing up Bermuda worse and worse every year; cahow tunnels were getting flooded or washed away completely. If they were going to survive, they needed to nest someplace higher and safer.

Madeiros tried an audacious but risky program: he took hatchlings from the mama and papa burrows on the low islands, moved them (but not the parents) into new burrows on Nonsuch, hand-raised them on fish and squid, and then prayed hard that once they left, in four years or so they would come back not to mom's original island, but to burrows on Nonsuch.

To help lure cahows to the new site, he played courtship calls through a solar-powered boombox. Madeiros had learned how to carry out this bait-and-switch scheme by apprenticing with the Gould's petrel folks in Australia. What worked successfully in the Pacific turned out to be equally successful in the Atlantic. There are now hundreds of cahows across all the sites, and the population increases yearly. File this news under *S*, for "Sometimes things work out."

An ostentation of peacocks, a charm of hummingbirds—nouns of assemblage give us the group names for animals, like a pride of lions or a murder of crows. So maybe it should be "a dice roll of petrels" or "a lucky break of cahows." They seemed gone, goner, gonest, and then, through luck and hard work, things turned around. As the climate spins hotter and hotter, and each politician seems more obtuse and self-serving than the one before, optimism may be hard to come by. Every nature lover has to ask the same question: Will the natural world survive us?

Or even more pointedly, given how dire things seem now, perhaps the real question is, do the stories of these multiple survivals suggest ways that we each can live in hope?

To quote Lady Anne in *Richard III*: "All men, I hope, live so."

WATCHING *PINGÜINOS*
AT THE REY JORGE DUMP

There are eighteen species of penguins worldwide, and most birders want to see all eighteen of them. Roger Tory Peterson, the bird artist and inventor of modern field guides, once confessed, "Penguins are habit forming. I am an addict." Jonathan Franzen, writing in the *New Yorker*, described visiting a chinstrap penguin colony in Antarctica. Even though the landscape was everywhere "smeared with nitric-smelling shit," Franzen was moved by the birds themselves. "Many of the adults had retreated uphill to molt, a process that involves standing still for several weeks, itchy and hungry, while new feathers push out old feathers. The patience of the molters, their silent endurance, was impossible not to admire in human terms."

In August 2011, a Humboldt penguin was discovered on Willoughby Rock, Washington state. This species is usually only found along the coast of South America. Birders wanted to know, was it self-sponsored and wild, or had it arrived as a stowaway on a fishing boat? If it was the first option, then it counted as a "first" on the official list of Washington avifauna, while the second possibility meant that it did not. It is sort of like asking if a hundred-dollar bill is authentic, or if it is a very good color Xerox. The second choice reduced the penguin to nothing more than an exotic pet or a zoo's out-of-pen escapee.

In order for the Washington penguin to be wild and not ship assisted, its presence so far north would need an explanation, and the most likely reason was that something had gone wrong with its internal compass. This happens more often than nonbirders might think. In an article for *Audubon* magazine, seabird expert Alvaro Jaramillo explains that "an experimental study on [off-route] blackpoll warblers in the Farallon Islands in the 1970s revealed that certain individuals flip their internal directions, causing them to confuse southwest with southeast." For this to make more sense, we have to remember that blackpoll warblers—the "poll" refers to their dark caps—are normally native to the Eastern US, not islands off of San Francisco. Jaramillo added that "the lead ornithologist, David DeSante, coined this phenomenon 'mirror-image misorientation.'"

This kind of mental reversal helps explain why California has so many warbler species on the official state list—forty-six New World species and five more Old World warblers. Some of those fifty-odd birds breed here; most do not. The rest arrived accidentally, after something went screwy and they flew in the wrong direction. In birding reports, the not-where-they-belong birds are called vagrants. It is not pejorative; birders like vagrants since they boost one's list. An off-route bird still counts the same as a native one, so long as it is wild and free-flying. My five hundredth state bird was a blackburnian warbler normally found in eastern forests, not Huntington Beach Central Park. Misfits and strays, you are my people.

Vagrancy also happens in seabirds. Washington's penguin could indeed have been a wild bird that accidentally swam north not south. It just happened to overshoot the equator by three thousand miles. There are precedents. As Jaramillo says, "Magellanic penguins are another intriguing case of vagrancy because they're highly migratory, with some moving hundreds of miles from their Patagonia breeding grounds to wintering areas as far as the upper end of Chile. Was it mirror-image misorientation that accounted for the penguin that showed up farther north on an El Salvador beach in 2007? Rare penguin sightings in the United States are always considered releases or ship-assisted transplants, but maybe they could be natural vagrants." South American penguins in Washington state are thus possible. "It's an absurd thought," Jaramillo says, "but then again, seabirds make the impossible seem possible."

Penguins can't fly, except they can, given how they use their wings underwater, which is with a vigorous and speedy rowing. Compared to the other birds, penguins have dense bones—dense like an ox's—which is useful for structural rigidity and useful because density reduces buoyancy, which helps when diving. According to one researcher, penguins swallow stones for the same reason. When not ingesting rocks, penguins eat sardines and sardine equivalents: anchovies, mackerel, gobies, crabs, krill, squid. It depends on their home range; krill is an Antarctic thing more than a Galápagos thing. I have a dim, misfocused memory of feeding sardines to penguins in a zoo as a child; the fish were cold and slimy and came in the same paper boat that french fries used to be served in. That is exactly how it happened, unless I have invented that entire scene based on a parallel, misfiled memory of feeding the zoo's sea lions.

Hominins have known about penguins for hundreds of thousands of years—for at least as long as we have known about elephants, zebras, pelicans, and hyenas. The species native to Africa used to be called black-footed penguin in English and was later named the jackass penguin, after its harsh, braying call. "Jackass" is a demeaning word in any culture and all the more so in the context of colonialism, so the species now is called the African penguin, or less often, the Cape penguin, after the Cape of Good Hope in South Africa.

According to most explorers, penguin meat tastes nasty. There are contrary views. A 1911 Japanese expedition to Antarctica claimed that the secret to eating penguins was to cook the skinned carcass with an abundance of miso paste, and that made it perfectly palatable. By this point in the journey, they had been away from home for two years, and in the wet hold of the ship, all the stored food had gone bad. In my view, the crew members were no longer reliable narrators. They probably thought shoe leather tasted fine.

If the penguin in Washington was indeed a wild stray, the question is not how it got there, but why we don't have even more errant penguins turning up along the North American coastline. Maybe every few years a wandering penguin does arrive in American waters, and we just never notice. As an example of how much we miss, there is a tiny seabird called Ainley's storm-petrel that was recently added to the California bird list. Fifty of these storm-petrels were radio-tagged on their nesting island off the coast of Mexico, and they checked in with the mama satellite every two hours, self-logging their foraging routes north and south of the US border. The data reveal that some were present in California during the winter of 2021–2022, but nobody had seen them. The species was here, but it came and went invisibly. We only know about it because of the auto-generated radio signals.

Mermaids, penguins, ghost ships, sea monsters: that we never see them does not mean they are not there. I experienced my first wild chinstrap penguins on a beach on King George Island, Antarctica. That is the island's name in English; I was staying with Chileans at their base and eating dinner with the Russians at theirs, and I was drinking grappa and watching porn with the Uruguayans at their station, so the people I was with called the island Rey Jorge. They also usually called the black-and-white seabirds *pingüinos*, as I still do today. I love the sound of the word itself. In my head, I always say it with an exclamation mark, ¡*pingüinos*!

The location was photogenic but fantastical, since that beach was below a former landfill, and there was a piece of white kitchenware—perhaps an old stove?—sticking out of the soil like whalebone. A blue-and-orange slick of leaking chemicals glossed the slope down to the tideline. It was a gray day but mild; snow was blowing sideways in small white dabs, little pellets of soft, snowy hail pinging against my parka shell, dainty and musical.

The *pingüinos* stood there, calm and dim, as the tide slurried some foam back and forth along the beach. The leaching chemicals painted the black sand with a pleasing rainbow sheen. Offshore a pod of minke whales broke the surface with abbreviated, bush-shaped blows, arched their backs, and dove out of sight one by one. A kelp gull circled the penguins twice and drifted out of sight. A skua passed us and did not stop. Just me and the drowsy penguins, sharing a quiet moment at a landfill. Torpid birds: desolate scenery: toxic leakage. Other than the bite from the snowy wind, it wasn't even cold.

"Reality is not the skin of appearance"—Simon Schama was talking about Francis Bacon when he said that, but the observation applies here too. Urban wildlife sightings can happen anywhere, even Antarctica. My best skua photograph from that trip happened at the fuel dump, and my first sighting of the all-white Antarctic snow petrel was over the airfield. Climate change means not only roller-coaster temperatures but also wilder weather patterns. Up will be down, and the impossible will be commonplace. There are eighteen species of penguins, and so far I have seen about half of them, some of them even on the parts of the planet where they correctly belong. The rest I am ready for now, no matter where they might turn up. I have my binoculars wiped clean and out of their case, and I am standing here, ready—so very, very ready.

ERNST MAYR WAS NEVER BORED

I would like to convince you to come with me to the Gulf of Aden to see sooty gulls and bottlenose dolphins and Jouanin's petrels, but if I were being fully honest, the motto for any given boat trip would have to be, "Come Join Us! It's Only Boring *Most* of the Time!" The fact is that most of the time, seabirding consists of hours of boredom interrupted by random moments of panic and coffee spillage. The drama lasts for a few minutes, then it all goes back to blank horizons and the same scruffy gulls trailing along in the wake, same as always.

In writing that down, I feel like I am revealing some kind of internal flaw. Admitting to being bored carries a vaguely shameful air, as if you're the one at fault, not the dull situation. John Berryman starts his poem "Dream Song 14" with this admission: "Life, friends, is boring." He hurries on to note that "we must not say so." It's poor manners to be bored, since after all, "the sky flashes, the great sea yearns, / [and] we ourselves flash and yearn." That may be so, but we nod off anyway.

You can see boredom spreading among participants on a pelagic trip the way spilled tea soaks into a tablecloth. Swaying side to side, befuddled by a double dose of Dramamine, worn out by an early start and a generous lunch, lulled into torpor by the white noise of the boat's *chugga chugga* engines, your eyes start to close and your head starts to sag when BAM, you snap back awake, sizzled into consciousness by a voice shouting, "Manx shearwater, Manx shearwater!" All the dream-dulled sleep zombies rise up at once and rush to the rail, binoculars ready, urgently pleading, "Where, *where*?"

False alarm. Not a true Manx after all—just a contrasty black-vented shearwater, a pretty enough bird (and globally rare), but you've already seen five hundred black-vented shearwaters so far today and can expect another five hundred before you get back to dock. Of those one thousand black-venteds, one per day might be a look-alike Manx, a vagrant in the Pacific. This is not their ocean; they belong elsewhere. Manx shearwaters are named for the Isle of Man, as in, the place in Great Britain, as in, they are native to the Atlantic not to the Pacific. There could be hidden breeding grounds in Chile, but most likely the Manx shearwaters seen on Monterey pelagic trips got blown in gales west past Patagonia, then once in this new ocean, they followed the Pacific side of South America up through the doldrums all the way to Baja and onward north.

Today's candidate Manx, whatever it was and wherever it has come from, no longer is in view. After a few minutes of desultory searching, maybe a visit to the loo, maybe a few sips of soda and a handful of crackers, everybody settles back on benches or stowage lockers (or, on the *Searcher* out of San Diego, folding lawn chairs), and struggles to scrutinize swell and sky. As the birders remind themselves, "You snooze, you lose." You have to keep looking, looking. It requires relentless attention. The best bird of the day could turn up any minute.

Or maybe not. Statistically, it is not likely. May as well try to rest. Worse than being sleepy is being drowsy but unable to sleep. You would *like* to nod off yet can't. Maybe you are seasick; maybe you are somebody who can't sleep sitting up. The boat chugs on, the sea stretches out vast and infinite, and nothing is happening. No rare birds circle the boat, and in fact, no birds at all: no seagulls, no shearwaters, not even some mats of kelp or the turtle-killing flotsam of a deflated birthday balloon, its Mylar skin silvering the surface of the sea.

It can be like that for hours, until the mind demands to be fed, yearning for some grand event. The ship hitting an uncharted reef, now *that* would be exciting. Instead, small patterns become surprisingly compelling. For instance, if we're chumming—trailing a line of fish guts and popcorn off the stern, hoping to attract a cloud of gulls—then stray bits of popcorn will get blown back on board and start to eddy across the deck, first this way, then that. I make a bet with myself which piece of soggy popcorn will get sucked out the scuppers first. A minor thought comes to me: Should I write down in my journal the suddenly remembered fact that in Spanish popcorn is called *palomitas*, little doves? By the middle of the afternoon, not able to sleep but too tired to be awake, I lack the willpower to initiate even that much writerly-ness.

Photography can be similarly too much to manage during my

deepest lethargy. If I do hoist the camera up, half the time I can't even bother to keep the horizon line level.

The boat's crew have their chores and petty crises (a minor fuel leak, a bathroom out of toilet paper, somebody's back-of-boat barf to hose off). Idle passengers can only stare and yawn, jealous of their drive and purpose. *The devil's name is dullness*—so said (so allegedly said) Robert E. Lee. How much of every day of our lives passes like this? Waiting at red lights, waiting for the kids to get out of school, waiting for an elevator, a flight, thick food in a slow microwave. So much waiting, so many pieces of time stolen from our total lifespan.

Is boredom a moral failing? I have been conditioned to agree with John Berryman's mother, who informed him that if he was bored, then he lacked Inner Resources, capital *I* and *R*. In the boredom poem, "Dream Song 14," she does not say this to him "repeatedly," which is the expected adverb. In the poem he claims she says it "re-peat-ING-ly," bringing to mind the first machine gun, also called the Gatling gun or the repeating rifle. *Dah-dah-dah-dah*: his mother hammers Berryman with synchronized barrels, each mother bullet followed immediately by another. Boredom as a moral failing—yes, this matches my own puritanical experience.

Not that I need my mother's voice to know I am a slacker. I never see enough, write enough, think enough, not compared to my heroes, anyway. Take biologist Ernst Mayr. His work on evolutionary theory causes people to mention him as a near equivalent to Darwin. Mayr lived to be 101, and after he retired from Harvard in 1975 at age 71, he published fourteen more books and two hundred more articles. Yet who was the person that Mayr most admired? Who was *his* hero? Charles Darwin of course, but also a seabird expert named Robert Cushman Murphy, who lived to nearly 100 as well. Murphy—for whom Murphy's petrel was named—ran the bird part of the American Museum of Natural History. Busy, busy guy. From the Special Collections website at Stony Brook: "The Robert Cushman Murphy Collection is comprised of 12 linear feet of correspondence, typescripts, photographs, negatives, notebooks, journals, book reviews, and slides." Twelve *feet* of materials, and that is just what is in one archive. He always was working. In trying to make sense of the Rothschild bird collection (which contained 280,000 specimens), Murphy had to create a handwritten and then retyped 740-page catalog and then pack a quarter of a million dead birds to ship from England to New York. That happened at the same time that he was running a national museum and peer-reviewing journal articles and keeping up on the literature and being a parent and writing up his own notes and mentoring young scientists like Ernst Mayr. I am sure he had help—secretaries, wives, colleagues, inner demons that drove him ever onward—but he also had old-fashioned elbow grease and drive. Clearly he was somebody who sat down and got on with it, day after day after day.

And then there is the case of his field notes. Ernst Mayr on Murphy: "With iron self-discipline, no matter how strenuous the day, he recorded his experiences in considerable detail in a diary, an extraordinarily valuable record considering the drastic changes in all of these places since then." What kinds of places? Antarctica,

Australia, Baja, the Caribbean, Ecuador, Fiji, the Western Mediterranean, Mexico, New Zealand, Olympic National Park, Panama, Peru, Venezuela—and I am sure I have missed a few. Murphy and Mayr come from a generation of thinkers who did their best science using language, not math. They had shotguns too, to collect birds for science, and as a young man, to get access to distant places, Murphy served on one of the last powered-by-sail whaling ships. But what mattered was writing it all down, while it was fresh in their minds, direct and uncontaminated.

These fellows are not bewhiskered fogeys from the antebellum past. We expect Thoreau and Emerson to have kept long, detailed journals; I think we sort of figure, "Well, back then, what else was there to do?" That's false, but even if it wasn't, Mayr died in 2005, and another Harvard luminary, E. O. Wilson (two-time winner of a Pulitzer Prize in nonfiction), just passed at the end of 2021. They are our contemporaries. Data mattered to them, but words did too, and as scientists, from field notes to best-selling books, they made sure to use language often and well. Who has time to be bored? There is too much work to do.

Ernst Mayr's most famous student was Stephen Jay Gould, another recently deceased Harvard glory boy. Gould argues that science arises out of boredom, or rather from a popcorn-counting refusal to be bored. I like him for that claim and for his willingness to quote from Karl Marx and American baseball stats in the same essay. The one time I met him, he gave a brilliant lecture even though he was coming down with the flu. The show must go on.

Gould tried to explain once how writing about geology in clear, insightful ways happens. In talking about geology, Gould was outlining art-making more generally. "Creative work," he writes, "in geology and anywhere else, is interaction and synthesis: half-baked ideas from a bar room, rocks in the field, chains of thought from lonely walks, numbers squeezed from rocks in a laboratory, numbers from a calculator riveted to a desk, fancy equipment usually malfunctioning on expensive ships, cheap equipment in the human cranium, arguments before a road cut."

According to this model, my problem on boats is not that I am bored, but that I am not bored enough. If no birds ever rose above the horizon ever again, then maybe I would be forced to do something else with my time . . . like write books, for example, to name one thing I try to put off. My friends admire my industry, but for any given book I finish, there were ten more I thought about and couldn't be bothered to start. Maybe to be more productive I just need more field trips to the road cut—more barstools and pocket notebooks—more Jules Verne (why don't giant cephalopods ever attack me? I have a sharp hatchet: I am ready)—more sharks (cf. Winslow Homer's *The Gulf Stream* at the Met, which has five sharks, eight flying fish, a water spout, and a demasted, rudderless ship)—more nicknames (even Gould himself called punctuated equilibrium, his celebrated theory of evolution, "punk eek")—more snarking and bitching (Mayr said that Gould stole the idea of punctuated equilibrium from him)—more cockroaches (which, according to Robert Murphy, helped control the bedbugs and other insects on board the whaling brig *Daisy*)—more mentors (from whom will we steal our ideas, otherwise?)—more popcorn, more chum, more gulls, more mermaids—more of everything, including more entries in my unfinished memoir about the radio stations we listened to when I was eighteen and what it felt like to be tan and lean at the pool edge, back when we were all young and perfect and we were never bored, because everything was still new, still possible, still about to happen, and besides, who could be bored when we were all flirting so hard it made the deck tiles curl? Everything was sunlit and perfect, and we knew, we just *knew*, it would last this way all the rest of our short, brilliant, never-bored lives.

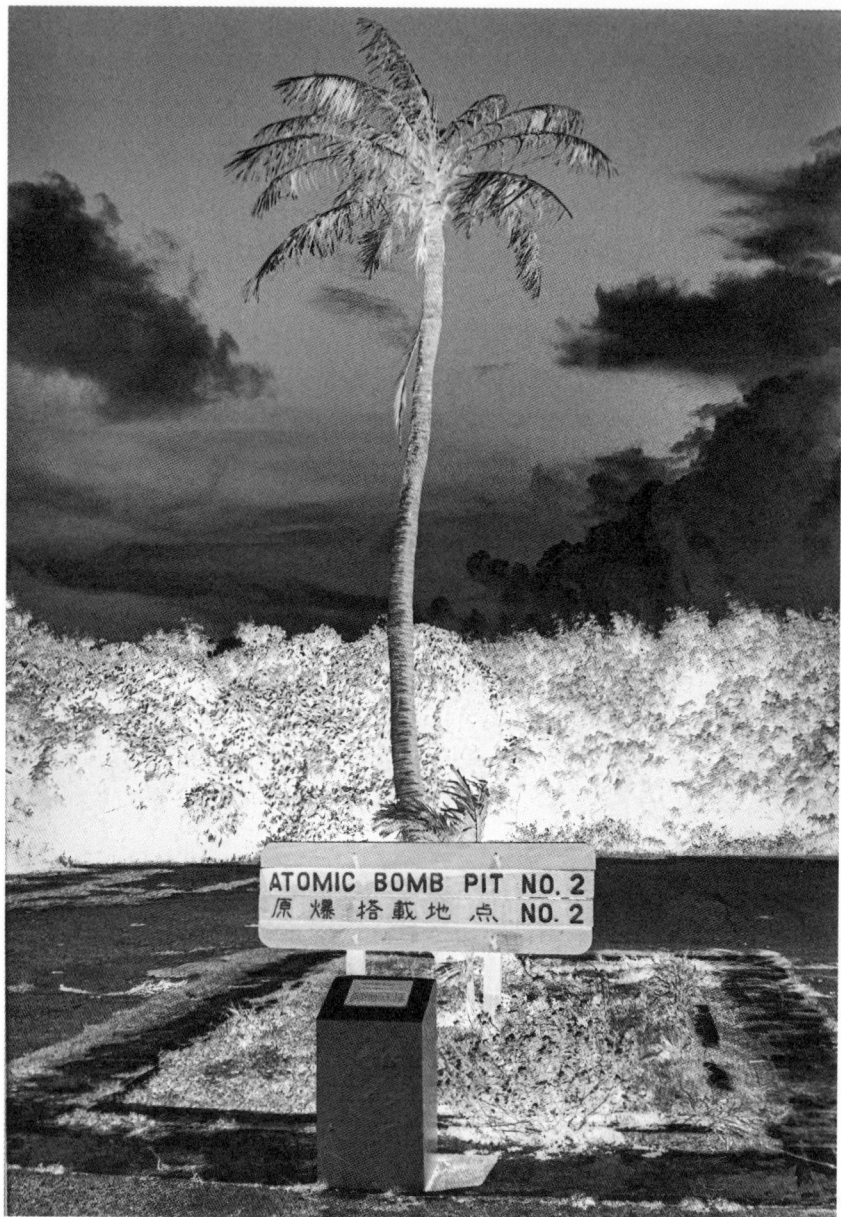

BOMBING THE BARNACLES

In your next life, if you had to come back as a you that was not you, there are worse options than being turned into a barnacle. To name one new attribute, a barnacle has the longest penis in relation to body size out of anything in the animal kingdom. It is also absurdly hardy; barnacles survived the atom bomb tests on Eniwetok Atoll. And deep-sea barnacles have been discovered five thousand feet under the ocean, clustered around hydrothermal vents. There is even a barnacle that thrives in that most toxic of sumps, the Salton Sea.

If asked to draw a barnacle, most people could come up with something approximately correct: a steep cone with some ostrich feathers poking out of the top. The feathers are really the legs, because barnacles are crustaceans, cousins to crabs and lobsters. Their legs, called cirri, comb the water, trapping plankton. Food is pulled inside the shell and digested. A barnacle does not have gills; the cirri handle gas exchange as well.

For such a small, drab lump, barnacles have a surprising number of fanboys. Charles Darwin liked barnacles, and in fact he liked them so much that he ended up publishing a thousand-page book about them. He did that when he very much should have been doing something else. For Darwin we might call the 1840s "the Great Stall." According to biographer David Quammen, during this time Darwin "continues fathering children, pottering around the house, acting like a hypochondriac; he dissects barnacles through a microscope and raises pigeons in a coop." Instead of doing all that, what he should have been doing is writing, revising, publishing, publicizing, and publicly defending the book that we know today as *On the Origin of Species*. That book did not come out until the final days of 1859, and only after a letter from upstart Alfred Russel Wallace forced Darwin to fish or cut bait.

One of the barnacle species that became known by way of Darwin's work is the striped barnacle, *Amphibalanus amphitrite*, first named in 1854. Native to the Indian Ocean, it is now found in every warm water port around the world. It's small—another name for it is acorn barnacle—and basically is a crusty thimble filled up with a gushy little crab-bug. Yet call it small but mighty, since it can flourish even in California's Salton Sea. This site was originally fresh water and so had no barnacles at all. The Salton Sea was formed by accident in 1905 when an irrigation canal failed, and the Colorado River flooded the Imperial Valley for two years. Since then, farm runoff and evaporation have made the Salton Sea twice the size of Lake Tahoe and twice as salty as seawater. It grows saltier by the day.

For the story of how barnacles got there, we turn to the Smithsonian's online compendium of invasive species, *Nemesis*. Here it says the barnacles were introduced to the Salton Sea during World War II, arriving from San Diego Bay attached to the buoys used to mark seaplane landing zones. Alternately, they might also have arrived in ballast tank water or attached to the hulls of pleasure boats, but if you ask me, the seaplane buoy needs to be the main way, for the essential reason that it is too good not to be true.

These days the shoreline of the Salton Sea can be white as coral sand, except it is not sand, it is a two-meter-thick berm of barnacle husks. Dried fish heads gape up as you walk, garnished by twisted driftwood and the lift and swirl of flies. Everything stinks, and in summer it is humid and yet 115 degrees. My late mother-in-law retired to Bombay Beach on the east shore of the Salton Sea, living in a silver Airstream under a white plywood awning. Every morning, she poured her coffee, waved at the neighbors, and practiced the French horn from eight to ten. After six months, it is a wonder she had any neighbors left at all.

During World War II, America rehearsed the Hiroshima mission using targets along the shore of the Salton Sea. The *Enola Gay* was not yet a named plane. It was one out of a dozen anonymous B-29s being readied for a secret mission. It was not until the night before the attack, at its base in the Pacific, that it was named after the pilot's mother, Enola Gay Tibbets.

To practice their daily bomb runs, each B-29 was loaded up with a concrete-filled dummy bomb. It filled the entire belly of the plane. These were all stand-ins, used to study technique. The real bombs were being invented two states to the east, at the Los Alamos Ranch School in New Mexico. They would be shipped to the South Pacific and united with their delivery planes later. The USS *Indianapolis*, the cruiser that ferried the Hiroshima bomb, was the one that was later sunk by a submarine and is cited in the movie *Jaws*, since so many of the crew were attacked by sharks. Almost 900 men went into the water when the ship sank; only 316 survived.

That was all later. First came the practice missions. Leaving their base in Wendover, the strike planes crossed the Utah and Nevada deserts and dropped their bombs on the barnacles of the Salton Sea, then turned around and flew back home. It all was routine, even boring. Last one to the bar was a rotten egg.

We think of barnacles as being tidepool animals, living in dense colonies on the sides of rocks along the shore in Monterey. The usual number cited is a density of seventy thousand individuals per square meter. To extrapolate out, that means two million barnacles would fit in an area covered by a skydiver's unfurled parachute. Barnacles are not our friends; among other problems, they clog the intake valves on boat engines. They also slow down cargo ships. (All barnacles interfere with the laminar flow of water, creating drag.) Each countermeasure has consequences, and the antifouling paint used to keep barnacles off hulls is nasty stuff. It would be better not to insert it into our already-overburdened waterways.

Yet it is hard not to admire barnacles. In the strange way that evolution always finds a way, some barnacles don't need a ship's hull or a whale's tail or a piece of floating wood to survive. Called the buoy barnacles, these contrarians are members of the neustonic ecosystem—an assemblage of plants and animals that specializes in living in the top three feet of the open ocean. Other community members include sargasso seaweed and a kind of daddy longlegs insect and the stinger-draped Portuguese man-o'-war. One specialist barnacle species joins these unanchored drifters. Instead of gluing itself to a rock or ship hull, the buoy barnacle creates a floating disc the size of a dime and then flips over, hanging from its own float. This anchor bobs on the surface and the barnacle dangles underneath, living upside-down for the rest of its life. It hangs there, filter feeding, content to drift wherever the wind takes it.

One day the name *Enola Gay* will mean nothing at all. The term "Hiroshima" will be as obscure to people as the Punic Wars are to us now, incidents half-remembered from late-night cram sessions. By the time that happens, the barnacles will still be there, unchanged and enduring. They will still be small, and still will be found worldwide, and still will be clogging intake valves and fouling ship bottoms.

Barnacles, for better or worse, will outlive us all.

AZAPPI

Sonno gli Soldati de Galera

Leopardo

PAGES FROM MY LAST FIELD GUIDE

When I am old and about to die, I want to write one last, final field guide, pouring into it all the nature that got cut from each of my previous books.

Like this detail, taken from my Svalbard journal: "A memory of a male walrus, very much in the mood, who self-pleasured himself with his own hind flippers."

The penis, I noted, was thick and red. In this last field guide, I also want to share how there used to be many kinds of walrus—twenty-three species in the fossil record for California alone. One of those was called the tuskless walrus. Poor thing. I mean, if anything defines "walrus-ness," it is the double-dagger tusks. A tuskless walrus is like a neckless giraffe: an animal best left to its fossilized fate.

Walruses in general are under-reported. When they turn up, it's in relation to a famous visitor or because they have sunk somebody's dory. News item: During a 2023 trip to Russia, North Korean leader Kim Jong-un was applauded by Misha the Walrus at the Primorsky Aquarium. The trained animal also blew him kisses. But such close contact does not always end well. In 2016 a man who took a selfie with an aquarium walrus in China was killed by the animal when it dragged him into the water. The animal's trainer, going in to save the first man, was killed as well. It is hard not to side with the walrus in this case.

Few field guides mention the essential truth that a walrus looks more like a hippo—especially when swimming—than it does like anything in the seal family. They forage with "piglike rooting" (direct quote from a mammal handbook) and appear "swollen" (same book). When cold, a walrus looks slate gray; when hot, they flush reddish pink. Surfacing, they give a little snorty blow, like an apprentice whale. Hawaiian chiefs carved neck pendants from walrus ivory, receiving the tusks from whalers bound for China with cargos of sea otter pelts. The Bishop Museum in Honolulu has one, and the Met in New York does too. They are both strung from cords made out of braided human hair.

Ask how do they "do it," and the answer is that walruses make love with structural assistance. In order to help it reach the necessary docking position, evolution has gifted the male walrus with a strut called a baculum, which can be three feet long. In a Native-owned shop in Nome I once bought such a baculum, clean and hefty, as a gift for my stepson Nick-Jack. He collected weapons—knives, staves, and swords, though never guns—and at the time, a penis bone war club seemed like something that was important for him to have. I was either the best parent or the worst, and my two sons could never decide which it was. That made three of us.

"They're like bloody starlings here"—said of vermiculated fishing-owls, at Tsam Tsam Lodge, Gabon. Also from my Gabon notes, from a campsite without electricity, one birder offering another a trade: "I will give you the last cookie for two AA batteries."

We were preindustrial for much of that trip, living on moonlight and matchsticks. Another thing I want to put in a field guide is how much wild mice covet candles, perhaps because they crave access to solidified fat. In my room in the research station, I slept under a mosquito net on the wooden floor, which was a perfectly comfortable nest except in the middle of the night, wood mice kept trying to steal my candles. They would snatch one and make off with it, carrying it crosswise in their mouths, only the candles wouldn't fit in the holes in the wall. Each morning when I got up, all my candles were still in the room, only they had been piled up in a logjam against the wall.

Gabon notebook: "The best part of my job is setting fires. Well, that and the leopards," a mandrill researcher from Belgium said to me. (He was using controlled burns as a way of regenerating woodlands.) An elephant once tried to kill him; malaria tried six times too. Mandrills are those large, extravagant baboons that have both red-and-blue faces and red-and-blue bottoms. Alan Dixson, in his book *The Mandrill: A Case of Extreme Sexual Selection*, says this about their courtship: "Males approaching females display a grin or silent bared-teeth face and make lip-smacks. This display may also occur with teeth-chattering." Does that sound like happy hour at the Yee-Haw Saloon to you? It does to me too.

There were villages in Gabon where all the young people, even the teachers, had moved to the capital city, Libreville. The only people left behind tending the overgrown fields and sagging, thatch-roofed huts were the eldest villagers, the ones who had not been seduced by the stories of all-night dance clubs and water fountains that emit nothing but a stream of the purest, coldest, freshest Coca-Cola. The abandoned hamlets reminded me of the social collapse after the first wave of the Black Death in the 1350s. Entire sections of European countryside had become rewilded by the decline of human population. Woodlands began to recover, and wolves reclaimed second-growth habitat. It was a strange resetting of the biological clock. Even near the heart of Paris, there were sections of landscape that were as tragic and wild as the Chernobyl Exclusion Zone today.

Thinking about Paris circa 1360 makes me wonder about European wolves more generally. For example, would wolves have been a problem for Chaucer's pilgrims on the way to Canterbury? Probably not; wolves seem to have been extirpated from central England by the 1300s, from Wales by the 1400s, from Highland Scotland by the 1500s, and from Ireland by the early- to mid-1700s. Culturally, of course, they live on: "a wolf in sheep's clothing," and all that. I used to think the wolf's livestock-snatching reputation was exaggerated, but I did once watch four wolves work over a sheep flock in Tibet, breaking the flock into attackable subunits as the stave-wielding guards ran back and forth, unable to keep up.

Eventually I want to write about the wolves of the California deserts, the subspecies known as the Mexican gray wolf or *lobo*. I

have seen this desert wolf in Big Bend National Park, long ago—
so long ago that in the same summer, I saw the last of the wild,
pre–captured and reintroduced condors circling over Los Padres
National Forest.

Meanwhile, in my hypothetical "I will write it one of these days"
guidebook to the Mojave, I want to consider the intersection between public land and military privilege. In Death Valley National
Park, you might think you're hiking in solitude when the sky is
suddenly torn in half by the low-level passage of fighter jets. They
are gone before you can look up, the sound echoing in rolling waves
for many minutes afterward.

What goes up sometimes comes down, and there are a *lot* of aviation wreck sites in the American deserts—"more than 600 in the
western Mojave alone," to quote research forwarded by the Center
for Land Use Interpretation. One story that I want to put in my
book is how in Death Valley, in a place informally known as Star
Wars Canyon (and officially known as "Rainbow Canyon"), Navy
and Air Force fighter jets make low-altitude runs through a narrow,
rocky gorge. In my guide to California mammals, I even feature a
picture of a cactus mouse my cousin took at that site.

More exactly: the jets *used* to make low-altitude flights in Star
Wars Canyon, past tense. On July 31, 2019, a Navy Super Hornet
missed a turn and crashed into the canyon wall at 600 mph, killing the pilot and injuring seven French tourists who were watching from the ground. In the official report, the deceased pilot is
referred to as the mishap pilot; he had a mishap wingman who
saw the fireball but not the actual crash. The F-18 was listed as the

mishap plane. I appreciate that is the verbal convention inside the aviation industry, but even so, the word seems overly polite, overly benign, given that the pilot's wife is now a widow, and the mishap airplane had originally cost $70 million. Perhaps that makes all the rest of us the mishap taxpayers.

As the Center for Land Use Interpretation points out, "these [desert] crash sites represent the meeting of the apogee of American technological sophistication, with the perigee of failure—the intersection of lofted ambition and terrestrial tragedy." The jets are still flying missions inside Death Valley, but they now make their zoom-and-boom runs through a canyon that is far enough into the park that the public cannot see what goes on. Crashes will still happen; next time it just won't be in front of civilians and their overactive smartphones.

As for why a national park was an aviation racetrack in the first place, it comes down to seniority. Access was grandfathered into the park bylaws. The jets can fly through the park in perpetuity, as low to the ground as that day's mission profile requires. It is their airspace; we only borrow it to go hiking or birdwatching.

Animals can suffer their own mishaps. In the melee that is a gannet feeding frenzy, diving seabirds birds do spear each other, sometimes fatally. Walruses stab each other while sparring, but it's not common; their tusks are large and obvious, so usually they can flash them like a mobster lifting his coat to reveal a pistol in the waistband, thus avoiding a confrontation in the first place. "Spat" (past tense of "spit," and also a little quarrel, so that is two meanings), but spats are also what gangsters wore over their shoes in speakeasies; "spat," plural noun, describes the larval form of barnacles. English—the gift that keeps on giving.

In my final book I will answer every question truthfully, starting with the most important: "At a hotel in India, how can you tell which guests are French?"

Answer: they are the well-dressed ones, smoking at breakfast.

India is and was exotic to the writers of Europe, as well it should be, given that "Solinus says that in India there are unicorns, elephants, and dragons in great numbers" (*Le Livre des Merveilles*, 1410). In my otherwise dragon-free journal I carefully wrote down that fact that in India, the Bishnoi tribal name for the gazelle is *chinkara*, which apparently means sneeze, after the animal's alarm call. "Look, crossing the road, there go three sneezes." "I like this mutton; it tastes just like sneeze."

Gazelles were the primary prey item for the once-abundant Indian cheetah. Akbar the Great perfected hunting with Indian cheetahs and reportedly owned a thousand of them. No wonder they later became extinct. (There is a reintroduction program going on now; the founder population has come from Namibia.) Physiologists note that the cheetah's spine is elastic and stretches when running, so that at sixty miles an hour, the cat maxes out to propel a seven-meter stride. Huge lungs, rudder tail, front legs connected to chest not by bone but ligaments, so the body hangs from a sling of muscles, ready to launch. The way their hip bones stick out, cheetahs look like fashion models from the nineties, the final examples of heroin chic.

It is hard to stay optimistic writing nature books in the Anthropocene. I did a book once about the best nature walks in California, and every time I picked a site and did my fieldwork, a week later it either burned down or a flood washed away the entrance road. I usually even had great pictures—the best pictures in the book—for the now-defunct site. Maybe I should have done a section, "Great places I wanted you to visit but now nobody can go there, because the planet is so out of whack and all." The plates would have looked *so* good.

In talking about what will be in my last book, that implies that I believe that as a format, books will still be around the rest of my lifetime, however short or long that is. I am not talking about e-books. I mean "book" in the literal sense of pictures and pages and covers. Books are sacred objects, made with care and love (usually) and treasured as such by readers. Certainly, we love our physical books more than we do blog posts or downloads. Each printed book is an individual manifestation of the idea of THE book, some master template of handmade perfection, and in talking about what I want in my final book, I really am talking about making all my previous books better by extending them infinitely. Each new book is a sequel to every book the writer has written before, and no matter how many times you rewrite the sentences before the final deadline, there is always one word that gets away. You see it disappearing into the dark like the taillights of the last bus, and you vow that next time—*next time*—you will work harder, think deeper, revise better.

Next time *for sure*, you will get every word right.

SEAWEED, STONES, SEED PEARLS, TWINE

When photography was brand new, nobody knew how to do it. One idea was to mix ferric ammonium citrate and potassium ferricyanide and coat a large sheet of paper with it, and then to take ferns, leaves, gloves, or seaweed—the physical objects themselves—and arrange them on the page. Put a pane of glass on top (to tamp it all flat) and set it in the sun for fifteen minutes. You now have a one-of-a-kind print called a cyanotype.

Once rinsed and preserved, these prints are a deep, saturated Prussian blue, except that the parts of the paper covered by plants or other master objects do not expose, so that once the specimen is lifted away, a delicate white outline remains. The silhouette combines graphic simplicity with scientific precision; the original plant is present through its absence. They also documented feathers this way, door keys, pressed flowers. By the standards of the 1840s, the process was shocking in its fidelity; the finest contemporary draftsman could not match the print's unerring accuracy.

The first master of cyanotypes was a botanical illustrator named Anna Atkins. She joined others in making it up as she went along. Pioneer photographer Hippolyte Bayard made botanical cyanotypes in France, while William Henry Fox Talbot, one of Atkins's mentors, experimented in Britain. Asterisk the word "probably" to this claim, but Anna Atkins does seem to have been the first woman photographer ever, as well as the author of the first photographic book.

In making her books, Atkins started by making prints of seaweed and moved on to ferns. The Museum of Modern Art describes her progress: "Freed from the imperatives of scientific accuracy, she focused increasingly on visual properties such as line and form, color and space, and transparency and opacity." While her contemporaries ultimately forgot about her, everybody praises her work now, including artist and seaweed wrangler Josie Iselin. "She was a hero to me," Iselin states, because she "was a polymath, a Victorian woman [who had been] schooled in the sciences of the day, including chemistry, to master the nascent cyanotype techniques. She was a superb draughtsman and naturalist." Atkins is now enjoying a revival; art giant Taschen recently released a deluxe edition of her work, and you can download her work from the Getty or order prints from the New York Public Library. Her work even appeared in a recent issue of *National Geographic* magazine.

Seaweed is a good subject for art. It is as old as the hills, or, depending where you're standing, older. Mt. Whitney, the highest point in the continental US, only started to rise up 180 million years ago, while the oldest-known fossil algae is over a *billion* years old. All seaweed species photosynthesize with every external surface, including their stalks. It is like a plant and yet not like a plant, since seaweed does not have leaves, flowers, or roots. The only concession to regular plant anatomy is that some seaweeds have a rootlike grabber called a holdfast, which helps anchor the central shaft. Best seaweed name: "dead man's fingers," though this term has also been applied to a wood-devouring fungus, to the blue sausage tree, and to a brand of rum.

Here is a secret about Anna Atkins: she had a gift for page design, no doubt about that, but before starting her photographic projects she must already have been noticing the strange and the marvelous—noticing and picking up this or that to save for later. Like me, like my friends, I think she was a scavenger of beauty. She found small treasures and set them aside. The edge of the ocean is often a good place to start one's searches. In the marine handbook *Between Pacific Tides*, Ed Ricketts observes that "one of the charms of strolling on a beach is the possibility that almost anything may turn up." He then engages in some dreaming of his own—today might be the day that he encounters "a case of scotch, perhaps, or some relic from the wreckage of a ship offshore." Quickly you dash forward. Once you reach it, though, "usually it will turn out to be a block of wood or piece of timber bored by *Teredo* [worms] and festooned with a dense growth goose barnacle."

To him that was good enough, as it is for me. So much to find, so much to do. String three shells and a silver bead on a twist of twine, and you have a bracelet. Or you might give a wave-glossed stone as a present, perhaps nested in moss. Back home from our excursions, we

line up our treasures on bookshelves and mantles and kitchen counters, or the bigger rocks end up in the garden or as one-of-a-kind
doorstops. Scaled up, we would have an emporium or a museum,
since a natural history museum, most critics agree, is a version of
the Renaissance "cabinet of curiosities." Scaled down, a single rock
is a mistake, trash, something that fell out of a daypack and should
be swept up. It is maintaining that balance point—more than one,
fewer than a thousand—that keeps the casual acquisition from edg-
ing into hoarding, or worse, something so grand and pretentious that
one now has started a collection. Collecting rarely ends well, since
even the wealthiest person alive cannot buy all the art or gold coins
or first editions in the world.

If you were to meet Anna Atkins today, or for that matter Emily Dickinson herself, what calling card might you hand over? In
my desk drawer, I have a mouse's tooth, salvaged ages ago from an
owl's pellet, small as the slimmest cog of a fine watch. Would that
be treasure enough? Shells, of course: a sea urchin's hull, sometimes
called a test, plus I have a delicate murex shell from a lagoon in
Madagascar, tines so sharp they pricked blood through my shirt. Or
one of my last acorns from F. Scott Fitzgerald's grave. A glass bead
from Carthage.

Most things I pick up do not have an immediate purpose. They
are cool, that's all, and they help me remember a good day or a special companion. Some items I will photograph and use for collages;
some I will make into jewelry or art. I do not know where nature
study will take me, only that it has not let me down yet. My job is to
be open to whatever comes next.

Two statements, equally true:

Every time I walk on the beach, I never find the case of scotch, the lost treasure.

And: Every time I walk on the beach, all I ever find is treasure, treasure, treasure, all the way down.

HOW TO PHOTOGRAPH A BIRD

Ah, good: here we have a seabird in the desert, one of my great joys.

And too, we have water in the desert, another joy.

Having gotten a shot that combines both at once, I have done my job. I can go home.

Or maybe not. We still need a caption. To be legible, an image needs to balance above a saucer of language. Images need words for the same reason a statue needs a plinth: without words the picture can't stand upright and be seen, be studied, be admired. Language completes the transaction that the image initiates.

The caption for this image should say something about the beast in question. We can see in this view an adult Sabine's gull with worn tail feathers making a banking turn in a medium-light rain over a freshwater marsh that is (though you can't see it or smell it) mostly comprised of tertiary-level former sewage. In the extended version, we might add that Sabine's gulls nest in Greenland, Canada, and along the top edge of Alaska. They mostly winter in tropical oceans and along the Humboldt Current of South America, and so in summer they are Arctic birds, and in winter they are seabirds. At no time of the year do they belong in the desert, but everybody makes a wrong turn once in a while; it is unusual but not unprecedented to come across an oceanic gull in the western Mojave. Oh, my snappy-looking but befuddled gull—how did you end up going so badly astray?

Captions intended to be studied by photography buffs usually add the camera maker (Nikon), the specific model number (Z9), the lens and aperture, the film stock (back in the day), and other numeric values that are pored over eagerly yet offer little true insight. It would be like asking a porn star what brand of condom he was using in scene twenty-one; the answer is a verifiable fact, but it is a fact without meaning. For our purposes, we can skip that datum.

"Photographs furnish evidence." Susan Sontag said that, though it's a truism that predates her by a hundred years. If I were to go to the bird committees to put in a sighting claim for a desert marsh Sabine's gull, I would need to upload a photograph to verify the record. A nice sketch would count too but would be suspicious in its departure from the norm. All birdwatchers have cameras now; it is part of the uniform. To claim something out of the ordinary requires proof. Yet we won't be able to rely on photographs as incontrovertible evidence for very much longer. Even with my limited Photoshop skills, I am nearly at the stage where I could put any bird in any habitat and end up with "proof." Just tell me the species you want to claim was in your backyard, and for ten dollars I will have it ready for you by close of business today.

This gull shot is a good picture but of course is incomplete. We can't see the parasites in this view; this bird probably has mites and lice. We can't see inside its head, but something skewed sideways for it to have ended up in the desert. This is a flawed individual, though not necessarily so flawed it won't survive. From here, if it keeps pushing south, it might make it to the Salton Sea and then on even more south to the Gulf of California (unless you are a fan of older maps, as I am, and still call it the Sea of Cortez). Or it could turn west to go over the San Gabriel Range to the Pacific Ocean, which is only sixty miles away. This is not a photograph of a bird—this is a photograph of evolution in progress: the eternal winnowing that separates the fit from the unfit.

"Fit" though means many things. Fit can mean not merely strongest and best, but the best suited—maybe this is the right fit for right here, right now. A related point is that sometimes being wrong works out, such as if this individual gull had discovered some magic bucket of infinite fish here in the desert, it might then come back year after year, and in future autumns maybe even bring its friends. They might set up a breeding colony here and save the long trek to and from the North Slope. Migration routes can change; you never know what piece of bad luck will work out to the good. In any given population, if a few birds stray off route, it could lead into a shift that benefits them all.

An odd example of an odd species either way, since Sabine's gull is "considered [to be] an aberrant gull, morphologically and behaviorally" (*Birds of the World*). General Sir Edward Sabine was an Arctic explorer and Anglo-Irish artilleryman who lived to be ninety-four.

There is a mountain named for him in Antarctica. Other "Sabine" birds and trees have been named after his brother, Joseph.

If this shot were printed in color, right away you would notice the bill tip, which looks as if it has been daubed with a quick lick of chrome yellow paint. The many shades of black on the feathers could be more easily distinguished too—the outer primaries are darker than Payne's gray but not as solid black as a chunk of charcoal. There is one paler feather in the middle of the lower back. That maybe is damage or maybe is a glitch in the DNA matrix, and so that feather is short X percentage of pigment. Photography worships specificity: no other Sabine's gull in the world has that missing smudge missing in just that way. A painter would not have captured it, and somebody preparing a taxidermy mount would have hidden it. Only this photograph at this moment documents that exact state of being. Photograph as eternity: I hope this bird likes being this age, this molt, this status, since now that I have taken the picture, it will be that way forever.

Things that I had to leave out include the wet vegetative smells, and the *scra-boom* of accelerating jets (this is, after all, an active military base), and the other birds in the sonic background, the grackles and red-winged blackbirds and marsh wrens, all surprisingly yakky for a rainy day in late summer. This shot shows a new marsh that was built on top of an old marsh. Precontact, Amargosa Creek flowed northeast through the Antelope Valley, ending up here, what is now called Piute Ponds. Seasonal pools would have been ringed with arroyo willow and cottonwoods. That river channel no longer exists, except after the largest rains; the groundwater is too low now for rivers, and houses and malls block the flow. Later there were duck hunting clubs and cabins. Now the marsh's ponds are diked in a rectilinear grid, and the water flow is controlled by culverts and gates. Other than when it rains, 95 percent of the water comes from treated sewage outflow. Nonnative tamarisks outnumber willows and cottonwoods ten to one. I have seen California voles here (twice), and I think there used to be muskrats, but if so, the coyotes got them all. There might be bobcats, but I never see them. Permit applicants have to take a desert tortoise safety training (protecting them from us, not the other way around), but I have not seen a tortoise here in many, many years.

The shot leaves out Native Americans—displaced from here by settlers and hunters—and all the aviation history, from stunt pilot Florence Pancho Barnes to the test pilots who as vigorous young men brimming with the Right Stuff turned into America's first astronauts.

The eBird checklist for Piute Ponds touts 319 official species, not counting things like escaped flamingos. If I ever do a Hockney-style photocollage, six feet by nine feet if I follow his *Pearblossom Highway* model, I want to feature a stray flamingo, and include as well the time I got stuck in the mud over my truck axles, and the time an untethered goose decoy fooled the heck out of me, and what a hundred white pelicans look like rising on a thermal, and the ways that northern harriers hunt like short-eared owls slowly quartering the fields, and sandhill cranes (rare but possible), and a confetti swirl of all the bar graph seagulls: western, herring, glaucous-winged, Bonaparte's, Franklin's, ring-billed, California, lesser black-backed, and of course, Sabine's, with its pulse of spring and fall records stuttering across the bar chart like bursts of Morse code.

My wall-filling photocollage of course will have Sabine's gulls in it—it *has* to, they are too compelling not to include. I would rather leave out all the swans and geese than leave those guys off. There are better places to go to see them, though. Mark this as the number one reason to go on fall boat trips off the West Coast of North America: to be able to see not only off-route singles but entire flocks of migrating Sabine's gulls, wings flashing black and white in the sun.

Well, that and maybe the blue whales.

You will need your best lens in that case; a Sabine's gull, at sea or in the desert, is not the kind of thing you will ever capture with an iPhone. Time to bring out the big glass, since these gulls fly fast and stay outside the reach of regular cameras. How to photograph a bird: delete everything you have ever shot and start over. Yes, you maybe were beginning to get good, but you can do better. You can *always* do better.

It's getting late—try to get some rest. It will be dawn soon, and before the sun comes up, we will get up, go out, and try again.

GOING TO THE MUSEUM
WITH WINSLOW HOMER

In the painting, a storm rages as an anonymous hero ziplines a damsel in distress to safety, using that decade's hottest tech, a rescue sling called a breeches buoy.

Wait a minute. Which painting?

Second floor, main building, Philadelphia Museum of Art: I have come to look with awe and irritation at Winslow Homer's great maritime painting, *The Life Line*, 1884. This storm-lashed drama marks a before and after for Homer, since this is the work that made him famous. Even the sales price was newsworthy, since it sold for $2,500 at a time when a bricklayer made $4 a day and you could buy a house for $1,000. (It was also five times more than Homer usually could get for his work.) If people could not see the original on display at the National Academy of Design in New York, they could experience it through prints. Not every critic loved *Life Line*, but nearly every critic talked about it.

This is sexy work, and then as now, sex sells. Homer knew it was going to be sexy when he was painting it, and the viewers knew it was sexy when they were looking at it. It transgresses with its very premise, since we have an unchaperoned woman alone in the arms of a strong, not-married-to-her man. The woman faces us, and her body is semiconscious, limp, open, exposed—a pose Homer borrowed from prior, more classical depictions of the female body (poses that were often fully nude). The rescuer's arms, wrapped firmly around her, meet with hands clasped at the bosom. At her throat, a touch of white must be the top edge of the shift covering her corset. Her wet dress clings tightly to her, and in fact, the dress has ridden up, revealing not just the hem of her petticoat—a wave-echoing sliver of white—but a peek at the tan flesh of her naked legs, visible as the circle of skin above her red garter belt and black stockings. Given the force of the storm, all this rent-garment, exposed-flesh drama is plausible, perhaps even artistically necessary under the requirements of realism. Pastors and deacons will not object.

To sex it up even more, Homer added a red shawl, which the gale is starting to unravel. A late addition, the wrap seems detached from her body and in fact spills into the scene like blood pouring from a wound. The fabric is thin, torn—a symbol perhaps for the fragility of her own life, and a visual chime with the flapping sails on the foundering wreck, but also a kind of two-degrees-of-separation metaphor for sexual surrender. Red flag, red danger: much is at stake here. And just in case we missed the symbolism, the red of the shawl is repeated above her knee, with our "avert your gaze" glimpse of a red garter.

More surprisingly—and this is a bit of mad genius on Homer's part—the blowing shawl completely covers the man's face. He is armored, as any good knight should be, since we can see the crown of his sou'wester hat, and that implies a full rainsuit. We can see he has heavy, square-toed boots too, heavier than he probably should be wearing in the water. But because of the shawl, his visor is down: tall, dark, and (presumably) handsome, he remains unknown to us. Did the crystal ball predict you would meet a mysterious stranger today? Perhaps he is the one. With the shawl's erasure, the man becomes agency without identity: he can rescue the maiden, take her to shore, tip his hat, say "Good day, ma'am," and move on unencumbered. "Who was that masked man?" the townspeople will ask themselves later.

Welcome to one of the greatest maritime paintings ever created, and it is one of ours, meaning it is by an American, which is all the more surprising since Britain was the nation that ruled the seas. We had really good whaling fleets, true, but this is not about that subject; Homer left that heavy lifting to Melville. In person, Winslow Homer eschewed drama; he was a small man who dressed carefully and well—like a banker, was the most common comparison—as if he was worried he would be evicted from the middle class if he did not dress the part.

How very American that the painting is conflicted about its own eroticism. Even as the work gives sex, it takes sex away: her legs are modestly crossed at the ankles, and legs, eyes, and lips are all closed to us. And not a vain person it seems: she is no spendthrift. Despite the storm damage, we can see that her dress is modest, almost drab. Winslow Homer knew his haute couture, and for this scene he has created a plain, middle-class, unfashionable girl. Contemporary viewers wondered if she was a schoolteacher or the captain's frugal daughter. She certainly could be somebody's sister, which helps keep the sexual tension from being more overt. Given her vulnerability, social values instruct us to protect her, not exploit her.

What a gloriously strange mess this painting is, with the whole greater than the mixed-up jumble of the parts. I love this painting and hate it, being troubled by the too-easy gender stereotypes and troubled too by how easily I am sucked in by the drama. As much as I pretend to be a cultural elitist, trained to appreciate the most astringent manifestations of modern art, the truth is I am as much a part of the peanut-crunching crowd as anybody else. Looking at this work, I am immediately and inextricably caught up by the action. Life balances in the momentary pause between swells. The rope sags under their weight, and everything is dire. They need to get to land, and soon—the woman, her clothes soaked through, shivers with hypothermia. She needs a doctor and warm blankets; she needs smelling salts, coffee, bread, love, prayer. The gentleman too probably needs a dry shirt and a shot of whiskey, and the chuck on the shoulder that says, "Job well done."

If Courbet had painted this same scene, the woman's skirt would have been slit past her bloomers, while if the scene had been painted by Thomas Eakins, oh my, what a fine face and strong torso the lifeboat captain would have. We all manifest sexuality in different ways, and for Winslow Homer the suggested *possibility* of an encounter takes things just far enough. That limit makes it sexier than anything explicit ever could be.

Meanwhile, what is going on with the dark cloak tangled around their collective legs and feet? She seems to be wearing a blue Batman cape, but if so, it's only drawn at half scale, and it starts somewhere around her waist and billows under their legs, being blown by a different angle of wind than the gusts animating the red scarf. What is the cape tied to, why isn't she wearing it over her shoulders, why does it look like a tablecloth for the children's table at Thanksgiving, and why hasn't it already blown away? The painting shrugs, unable (or unwilling) to explain. It is there because it needs to be there: the painting looks better with it, no other reason.

Contemporary audiences responded immediately. What to us seems like over-the-top melodrama read to the people of the time as sincere and true, since when this was painted, almost everyone knew somebody who had died at sea. In one month alone, January 1884, there were over two hundred shipwrecks in Europe and North America. Ships routinely went aground; they blew off course; they smashed into each other; they sank without a trace— by the late nineteenth century, shipwrecks were as common as midwinter flight cancelations in Chicago are now, only with worse consequences. Death by drowning felt inevitable; if it was not going to happen to you, then law of averages, it was going to happen to somebody you knew.

By the time Homer made this work, the federal government had created a new Life-Saving Service. It centered on deploying the breeches buoy. Simple enough apparatus: a cannon shoots a line from shore to ship, and that first line is used to set up a sturdier line, and then shore-based rescue crews send across a cork life ring with a pair of canvas trousers attached. That hangs from a pulley. Step in, put your legs the rest of the way through, and hang on as a rope attached to the pulley is hauled by the rescue team to bring you back on shore. It was intended to be a solo ride, but if your state of mind was such that you couldn't step into the pants-slash-ring, then a rescuer could insert themselves instead and hold you as you both rode over (and through) the water together. The breeches buoy system lives on, though it has been modified to hold rescue crews winched down from a helicopter. The painting's "life line" title extends a Christian reading onto the scene. One hymn stands in for many. Edward Ufford's "Throw Out the Life Line," 1888, includes this verse: "This is the life line, oh, grasp it today! / See, you are recklessly drifting away; / Voices in warning, shout over the wave, / O grasp the strong life line, for Jesus can save."

A few critics grumped at Homer because his water wasn't watery enough. *Oh, come on*—it's fine, especially in its loosest, most impressionist passages. Art historian Robert Hughes has singled Homer out exactly because of his water, reminding us that "no artist since Turner had painted the sea with such lyric concentration, from the beaming blue transparency of the Caribbean, captured in masterly watercolors, to the sullen beat and topple of gray combers driven by an Atlantic gale on the Maine rocks." Homer's oceans are pieces of ecology that can kill you, that's the main lesson, followed by the reminder, "Always check your knots."

According to this painting, when it comes down to man versus nature, in the end, man will win. In this sense, "man" as in "mankind," but given the gender binaries being asserted by the work,

Homer also means man as in the embodiment of the masculinized gender. His claims are meant to be reassuring, even if Homer implies that it all will be hard work. We will need sturdy hawsers and even sturdier men, but with enough bravery and social cooperation, any storm can be weathered and any chasm crossed.

I am almost persuaded. A movie that updates this painting is *Captain Phillips*, 2013, starring Tom Hanks. In the film, the kindly and inoffensive Captain Phillips is kidnapped by Somali pirates. Quickly we realize that in this equation, Tom Hanks is the maiden in need of rescue; he will show cunning and fortitude, but outside agency needs to intervene if he is to be saved. We experience much drama over the two-hour runtime, but in the end, a beaten and abused Tom Hanks is saved because of the steady nerves and superhuman marksmanship of Seal Team Six. From the fantail of the USS *Bainbridge*, at the last possible moment, good guys shoot bad guys, and all is right with the world.

I missed this movie in the theaters but clicked on it during a long flight home from Asia, and once in, I was captivated. I watched it over and over, mesmerized by the sangfroid of the rescue team and enchanted by the kind professionalism of the navy medics who treat Tom Hanks at the end. It was Winslow Homer all over again, but with night-vision goggles and sniper rifles, the rescue tech of the modern era. The painting lives on, except it now comes with a fifty-million-dollar budget and modern-day pirates.

At the same time that it reassures us and reinforces a traditional narrative, Homer's painting asks questions it refuses to answer. If I want to know why the rescued woman is both conscious (upper hand on rope, legs intentionally crossed) and yet simultaneously un-conscious (head lolling, body slumped), too bad for me, since the painting never resolves the mix of alert and succumbed states. The same if I want to know where the light is coming from—how can we, the viewers, "see" this scene?—there's no easy, obvious answer to that either. Homer has invented an omniscient point of view sim-ilar to how nineteenth-century novelists created their own all-seeing narrators, and invented (preinvented) the omnidirectional China ball glow of studio lighting used in Hollywood from the 1930s until now. The scene just exists—it just "is"—and never mind where the light is coming from. If you have time to ask questions like that, the painting implies, then go put some coffee on the stove and stack up dry blankets. Make yourself useful for a change.

It can be hard to see good art. Hard literally—some museums are far away; some keep frugal hours—and hard metaphorically, since some artists are hidden under so many layers of fame and public-ity that looking at their work is like trying to peer into the tinted windows of a bulletproof car. "Hello?" *Tap, tap, tap.* "Anybody in there?" Winslow Homer was famous enough to have been collected broadly in his own lifetime and yet is now forgotten enough that you can linger all day and not be in anybody's way. That makes him an ideal artist: accessible but ignored. Of course, pity the painting that has to share a wall with him; he is, as the expression goes, a hard act to follow. Who wants to look at a vase of flowers or a field of wheat when the entire Atlantic Ocean is surging back and forth, ready to bust open the gilt-wood frame and pour onto the floor? I am not even certain I am ready to look at it myself, and I'm ten feet back and wearing sensible shoes.

We go to museums for many reasons, including to find a quiet place for tea and cakes or, in my case, to hit up the gift shop for discontinued, half-price postcards. *The Life Line* reminds us of a final, better reason: we go deep into art to know what it feels like to die the worst of all possible deaths, and then, a moment later, to be saved and carried back to life, back to family, back to the perpetual safety of the firm and welcoming shore.

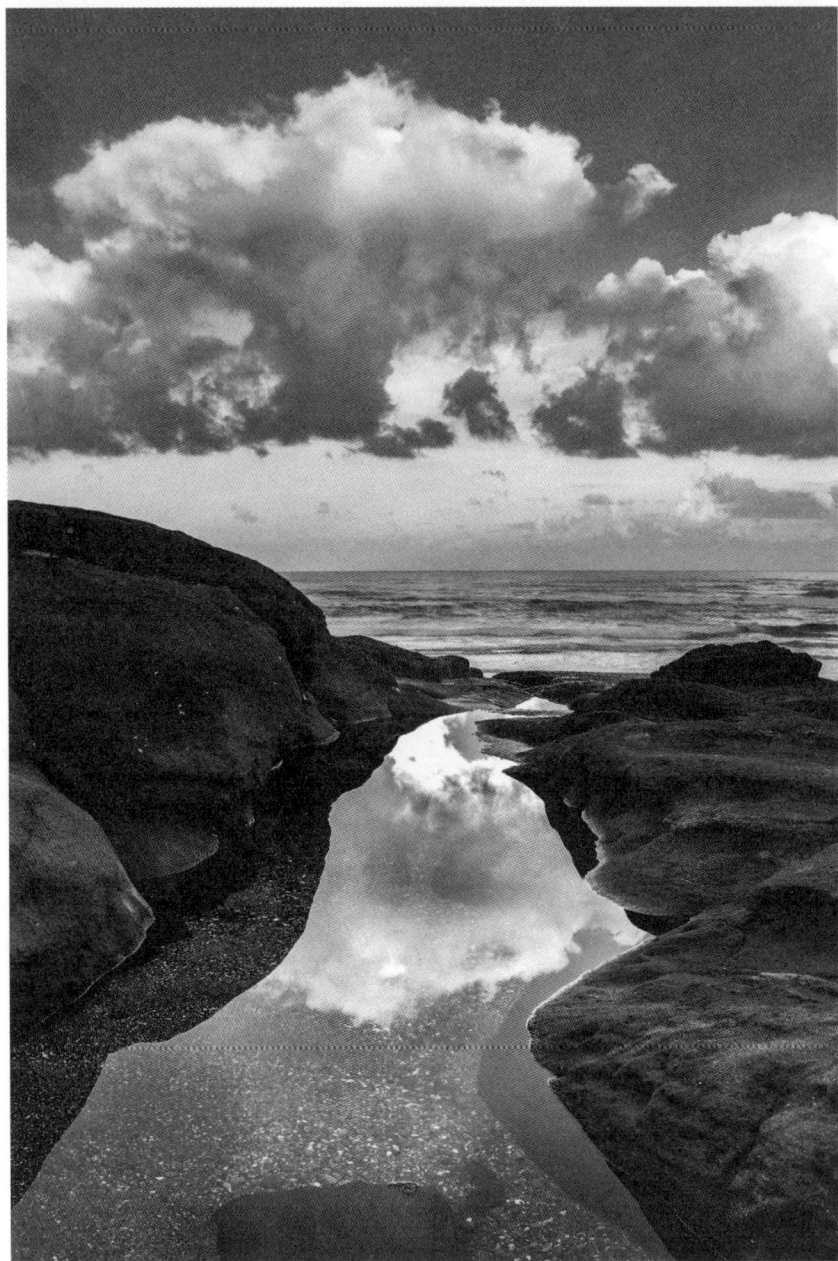

THE BIBLICAL TIDE POOLS
OF CANNERY ROW

Inside of every lunging, churning, road-flooding high tide is a nice, calm, helpful low tide trying to get out.

The very lowest of low tides are called minus tides, so named because tide charts list their values as negative numbers. When minus tides happen, glory praise to Neptune, since now we experience firsthand the least harvested, most abundantly inhabited, very best tide pools of all. The sea has taken away the water and left behind the starfish and sea anemones and eel grass, making undersea explorers of us all, no dive masks required. Tide "pools" since the rock naturally holds miniature lagoons and intricate passageways; the pools may only be six inches deep, but magical worlds wait for us inside each one. Tiny, darting fish zip shadow to shadow in clear water, while sea anemones wave slowly, their soft tentacles the green blossoms of pulsing death. Hermit crabs transform seashells into ambulatory castles, and my imagination wants every black slip of kelp to hide a skulking octopus.

All minus tides are urgent experiences since you only have an hour or two until the sea begins to return. A hat is a given, but I vacillate in picking ideal footwear for my excursions. I like the traction and "who cares if I get wet" part of wearing rafting sandals, but rocks can be sharp and barnacles sharper still. Infected cuts can take a long time to heal, so for me, rubber boots usually win out. Polarized sunglasses help banish glare even on overcast days. Macro lenses for the camera, yes of course, and I can't decide if a trekking pole helps or if it gets in the way. Most essential of all, though, either in the bottom of my field bag or else waiting in the car, will be the book—THE book, *the* book, the most important tide pool book of all: *Between Pacific Tides* by Ed Ricketts.

Ed Ricketts, the hard-partying "Doc" of Steinbeck's *Cannery Row*, probably would be surprised to know that he has research boats named after him and statues raised in his honor. Steinbeck in the novel describes Ricketts as deceptively small yet tough and wiry, with a bearded face that was "half Christ and half satyr." He had the hands of a brain surgeon and tipped his hat to dogs. Ed Ricketts loved books even more than I do. In 1936, a fire that started in an adjacent cannery destroyed most of his laboratory. Because of a pending divorce, Ricketts had also been living there at the time. He lost his specimens, his notes, his clothes, most of his manuscripts, and all of his personal library. Insurance only covered a quarter of the loss. That winter his friends helped him rebuild his collection, and at his request, they replaced his poetry and philosophy books first, one Christmas present at a time.

Ricketts, a college dropout, might be especially shocked to learn that his marine ecology handbook, *Between Pacific Tides*, still remains in print. First published in 1939, this book represents "one of the classic works of marine biology" (to quote the Stanford University Press website) and is a book that has been "a favorite for generations." In the book's foreword, Steinbeck equates the author to Aristotle and Pliny.

One reason that *Between Pacific Tides* matters so much is that where others saw scenery, Ricketts saw habitat, and habitat, Ricketts realized, is always subdivided in subtle but indelible ways. At the time of publication, what is now an ecological commonplace—the reality of microhabitats and niche partitioning—was a radically new way to present natural history. Nothing along the shore to him was ever simply "old pieces of wood" or "wet mossy rocks." His book trains us to see life zones such as sand flats or wharf pilings. In fact, wharf pilings are so interesting and varied, in the Ed Ricketts universe there are five different kinds of wharf pile habitat. Once you know how to read the shorescape, the ocean's edge becomes even more surprising and abundant.

Other handbooks focused on keying out specimens but provided little or no ecological context. And other books were not as nearly as ambitious as this one. Aided by a no-drama, mostly silent coauthor, Jack Calvin, Ricketts does his best to know everything, share everything, show everything, from the lowest of low tidelines up to the highest, most exposed rocks. My edition, many years old now, spans 652 pages. From Zuma Beach to abalone, it's all there.

Ed Ricketts was born in 1897, which makes Amelia Earhart, William Faulkner, and Thornton Wilder his contemporaries. In his own lifetime, he was often admired and befriended, but at the same time, he was often broke and misunderstood. Among other unpopular and pre-trend insights, he was the first ecologist to predict that with continued overfishing, Monterey's sardine population was about to crash. Nobody listened, and not only did it crash, but the collapse was even worse than he had predicted. Today of course he would be frantically waving a climate change flag and trying to get coastal cities to plan for the inevitable. Everybody likes to watch big wave surfers, but nobody wants to have to watch them from their roof as the waves eat up the lower half of the house.

Ed Ricketts died in 1948 at the intersection of Drake and Wave, in the middle of what is now the tourist version of Cannery Row. "Drake" as in Sir Francis Drake, the Tudor pirate, credited as the first Englishman to see the Pacific. By morbid coincidence, my favorite motel overlooks this same corner. Ricketts was coming back from the store when his secondhand Buick stalled on the tracks, and his car was hit by a train. He died three days later.

His papers are archived over the hills and up north at Stanford. These reveal the hard work of making a living as a naturalist. For example, in box one, folder nine, researchers can review his correspondence with Walt Disney Studios, which ordered slides of amoebas in 1939. Ricketts revered life, but his day job involved killing nature, putting it in jars, and selling it to universities and high schools.

As a good and careful scientist, Ricketts wrote daily notes and kept separate index cards for every species he encountered. To

scrape barnacles from the backs of whales, Ricketts went to Moss Landing, where there was still a shore-based whaling station. The flensed carcasses smelled terrible, and the rancid oil ruined his favorite hat, but he did what he had to do to ensure he had robust data.

At the same time, raw data by itself was never enough. His oceans churn and crash; ships sink, and shellfish can change their sex. He likes hermit crabs and says so. In writing one of the best and most comprehensive field guides ever, Ed Ricketts and Jack Calvin also stop and smell the sand verbena, or at least stop to cook up a quick beachside dinner. Owl limpets are territorial gastropods common on most Pacific Coast rocky shores; in her illustrated guide, Marni Fylling calls them "one of the few farmers of the tide pool." They graze algae, which they cultivate and defend against interlopers. Limpets score striations as they graze, leaving rock surfaces finely corrugated. (The "owl" part comes from a lining inside the shell, which looks like an owl's face.) Ricketts logs the usual natural history features, but in his entry, he adds other small, vivid, essential details, such as the fact that "Mexicans justly prize the owl limpet as food. When properly prepared, it is delicious, having finer meat and a more delicate flavor than abalone. Each animal provides one steak the size of a silver dollar, which must be pounded between two blocks of wood before it is rolled in egg and flour and fried." Wait a minute—where did *that* detour come from? It is as if a Jacques Cousteau TV special has suddenly been hijacked by a slightly drunk Anthony Bourdain.

Ricketts lived on the Monterey coast when that place was home to Robinson Jeffers at the early end of the timeline and Henry

Miller at the far end. Plein air painters centered in Carmel had created a Bohemian colony sometimes called the "Elysium of the Pacific." During this same period, photography shifted from soft-focus pictorialism to the clean, hard modernism of the f/64 movement, and Edward Weston made some of his most iconic photographs at Point Lobos. Weston and Ricketts attended each other's parties, drinking cheap wine and arguing about Carl Jung long into the night. Those parties were the place to be according to Steinbeck, his fellow author Joseph Campbell, and artists of all kinds. In an odd footnote to the twentieth century, Xenia Kashevaroff, a surrealist artist of mixed Russian and Native Alaskan heritage, and her then-husband, the avant-garde musician John Cage, helped Ricketts index the manuscript version of *Between Pacific Tides* during a Christmas visit to Monterey. The three of them listened to Stravinsky and argued about James Joyce.

Since then, Monterey has changed. Steinbeck opens *Cannery Row* with this sentence: "Cannery Row in Monterey in California is a poem, a stink, a grating noise, a quality of light, a tone, a habit, a nostalgia, a dream." He then pans the camera to take in Cannery Row's bars and houses of ill fame, its corner grocery stores, and, plural, its marine laboratories. Cannery Row today is a touristy mess of e-bikes and fudge factories, bickering gulls and tired children, overbooked restaurants and underlit parking garages. You can stand in line half an hour waiting to buy tickets to the aquarium. The stench and bustle of sardine canneries is long gone.

Coastal nature has changed too. While the sea otters have come back, other things have disappeared. When Ed Ricketts explored

Monterey pools, he found an octopus under every fourth rock. I mean that not as metaphor but as fact; we know the success rate, since those who went collecting with him said as much. His tidelands held tens of thousands of sea stars and many millions of mussels. When Ricketts was exploring the coast, the oceans were full, and the beaches were not. You look at old photographs of a lone structure like the lighthouse at Point Pinos, and it's all just so . . . so . . . miraculously empty. No golf course abuts it; no crows shred lunch bags; there is no evidence of classic car weekends, when the hotels will pass 100 percent occupancy. To a naturalist it must have seemed like paradise, and apparently to real estate developers too.

We can get a sense of that former abundance during the lowest tides of the year, when the ocean gives us back the protected coves and lost grottos. For me nature is always tangled up in memory and culture, childhood and reading. In the firmly administered doses of Protestant Christianity that I received every Sunday, I was encouraged to see the Bible as true history. Its miracles really did happen; real people did and said the things we were learning about in that week's lesson. One part of the Bible that I did spend a lot of time thinking about was when Moses led the Israelites out of slavery in Egypt, in the book of Exodus. In that tradition, Moses (with God's help) parted the Red Sea, and afterward, the sea closed behind them and killed their enemies.

I liked this part, not the dead enemies, but how the story kept the water at a comfortable distance. The text emphasizes how the fleeing Israelites walked on dry land. I did wonder about the specifics. How dry was it? I decided it was firm underfoot but damp; the heaviest patriarchs left sandal tracks indented in the fine mud, their footprints slowly filling with water as they rushed through the temporary gap. As I knew from California beaches, footprints can last all day until the rising tide erases them. Scattered among the silver ovals of the footprints might have been household items like dropped lamps or fascines of kindling.

The ocean itself was divided on either side into two vertical walls of water, fifty feet tall—or at least it was that tall in my imagined version. In that innocent and preinternet time, I never thought to ask myself, "So how deep was the Red Sea, anyway?" Today it supports dive resorts and military bases and oil wells; this is contested water, deep water. You can read your email today because of the Red Sea: it is bisected by multiple subsea cables. An estimated 90 percent of communications between Europe and Asia and 17 percent of global internet traffic traverse cables under the fourteen-mile-wide Bab al Mandab Strait. The data streak from one relay to another at two-thirds the speed of light.

As a kid, none of that was on my mind, or on anybody else's either. To me the biblical site was something bigger than a lake but not so tall as the sky. Fifty vertical feet of water—in my mind, that should about do it.

I knew what to expect from the seabed because even on average beaches, average weekends, back then one could still find diverse tidelands and open shores. My family liked to go to Laguna Beach for the day, a long drive from East L.A., but it took us to a classy place; one Sunday night I came home with not one abalone shell, but two. These were in shallow water, free for the taking. We never could have imagined they would ever run out. I also came back most Sundays with a scarlet sunburn—is this the origin story for my recurrent melanoma? My mother had made a picnic before we left, and my dad had filled the big jug with a slurry of orange juice, ice chips, and Seven Up. Even now I am nostalgic for that taste, though I prefer to cut the sweetness with a slug of vodka and a tumble of quartered lemons.

When there was enough money, we took the train to San Diego and stayed in a motel facing the water on Pacific Beach. There were mussel shells here and scallops and razor clams, the same as at Laguna, but the real prizes were the sand dollars, many of them still intact. There must have been hundreds. My mother warned me to rinse them in the sink to make sure all the meat inside was scooped out, or else after a few days they would make my shell bucket stink. Good advice then; good advice now.

At the time I never wondered what sand dollars looked like underwater. I always assumed gray, beige, and tan were the default states. When I took marine biology in college, the course was held at a field station in Baja. Wearing a life jacket and keeping my fear on the tightest of chokeholds, I frantically paddled out from shore each day to complete my homework. I was embarrassed to realize that in the wild, living sand dollars were purple and slightly fuzzy. (Ricketts compares them to red velvet.) It made sense they should be different when alive than what they looked like dead and washed up on the beach. I had never thought about it before.

Later, checking my copy of *Between Pacific Tides*, I learned that sand dollars rest flat at low tide, turn sideways like embedded Frisbees to filter feed in moderate currents, and then when the water becomes too rough, go back to hunkering flat once more.

Each of these trips to the beach helped me better visualize my Bible lessons. Learning about Moses, during the Red Sea part of the story I imagined the ocean's walls of water as transparent and quivery, like clear, thick, Jell-O, and I assumed that you could thrust your arm into it, reaching through the gelatinous boundary layer and into the liquid water behind. I knew what Jell-O felt like because we ate it for dinner two or three times a week. I imagined that if you put your face right up against the edge of the water, you would even see fish. These schools would edge up to the wet-dry boundary, see all the robed people walking on our side of the divide, and swing back in a panic, darting away. For them, too, the day would have contained miracles.

Best thing of all, the seabed must have been covered by seashells. So many shells! If I could have been there, I shivered to think of everything I could have found. I knew for sure I would have been scolded for falling behind. Pharaoh's army or no Pharaoh's army, no way was I going to pass up so many great shells. I would have needed a satchel to carry my treasures, or perhaps I could have made some kind of apron out of my robes. What if there were old spear points, or bright slivers of coral? How sad I would have been when the far side was reached and Moses commanded the waters to close behind him, shutting the ocean once again.

Exodus of course was never supposed to be about nature; my proclivities have made it so. And yet my remembered seashells are nearly as good as the real ones that have come into my life since then. If you hold a seashell to your ear—and it needs to be a large one, clean and smooth, from the part of the beach that's so deep it is mostly ocean for most of the time—you can hear not only the sea in it, but you can hear the Bible lessons from when you were five years old, and seagulls and Coppertone ads and lip balm and sandwich crusts, and your mother's voice, and you shouting back—shouting to Mom or God or to Moses—"No, no, not yet. It's too soon to go. I'm not done yet. Just ten more minutes—*please*."

MELVILLE IN PATAGONIA

1.

What made Melville, Melville?

The formula would have to include the way he was in love with Nathaniel Hawthorne but couldn't admit it, even to himself, though again, with Hawthorne, have you ever read his letters? Seen his portrait? As in, jeepers, who *wouldn't* be in love with him? Your take may differ, but in my reading of the triangle, Sophia Hawthorne, singular, somehow lets it be known to Melville—a conflicted man, bipolar, a great writer, but what do we do with the hundreds of his pages (such as all of the poem *Clarel*) that are now utterly unreadable?—that if he tries to seduce her husband again, she will stab out his eyes with hat pins and leave him to die at the side of the road. We could even argue that Sophia was the more indelible writer of the three, since in Concord, when Sophia and Mr. Hawthorne had just been married, she used her diamond ring to etch this in the window of the Old Manse: *Man's accidents are*

God's purposes. Sophia A. Hawthorne. It is still there today, and it was not even their house—it was a rental.

When traveling, Melville recorded his thoughts in pocket notebooks. So far as we know, most or all of them have survived. You can click through digital versions online. Melville's entries vary from the quotidian—his laundry will include his smalls and his nightshirt—to the acerbic. He was in an especially bad mood in Jerusalem, winter of 1856–1857. By then the publication of *Moby-Dick* had come and gone, his chance to date Hawthorne had come and gone, and Melville (probably due to marital trouble at home) had been sponsored by his father-in-law to get the hell out of Dodge, in this case by touring the Holy Land. This was before the development of mainstream biblical tourism; going to Palestine was still a rough and risky journey. To make his shekels last, Melville traveled second class.

His dark mood followed him from America to the Levant. Judea, Melville wrote, was "a tornado of stones"; the only place of interest to him was the graveyard. Typical notebook entry (with original dashes and spelling retained): "The Holy Sepulchre—ruined dome—confused & half-ruinous pile.—Laberithys & terraces of mouldy grottos, tombs, & shrines. Smells like a dead-house, dingy light." Melville mocks the guards and the tourists, the food and the hotels, the shills and the marks. Just as irksome, he was often pressed on the quality of his faith. "Do you believe, Sir?," pestered a missionary at lunch. Reflecting on the region as a whole, Melville wonders, "Is the desolation the result of the fatal embrace of the Deity?" If so, "Hapless are the favorites of heaven."

In entering thoughts in the notebook, Melville could be so compressed that his entries turn him into an honorary Imagiste. (Ezra Pound, about abstractions: "The natural object is always the adequate symbol.") Lineated, Melville's prose even looks like modern poetry:

On each stone doorstep
the eyes of the children closed by sores
& over it all
the feeding crawl of flies.

His observation implies moral action: if this is how the world is, then one must witness and testify, documenting the world to change the world. One wonders if in modern times Herman Melville might have been a good war photographer, going in with Leica and flak jacket to document Gaza or Ukraine. He did not mind traveling alone, and he always needed the money. Doing that kind of frontline journalism could well have reinforced his skepticism, leaving him even more melancholy and isolated. He might not have minded. Is the world a cruel and random place? Ah yes, he suspected as much.

2.

Melville came back from the Holy Land intact, if not healed—it survived him, and he survived it—and then a few years later, after a failed circuit as a public lecturer, he went away again. The 1860 notebook finds Melville on board the *Meteor*, a clipper going from Boston to San Francisco by way of Patagonia. This was forty years before the opening of the Panama Canal, so they have to make it the long way around, sailing against the wind past Cape Horn. (Cape Horn is the place Ishmael lists as his goal when he joins the *Pequod* in New Bedford.) Melville had been here twenty years before, outbound on the whaling ship *Acushnet*, and then again a few years after that, coming homeward from the Pacific as a deckhand on the frigate USS *United States*. Whatever he felt about crossing the same perilous piece of ocean in the 1860 trip, he had no choice, really; the transcontinental railroad was still a decade away, and the land crossing in Central America had its own hazards.

Even now Cape Horn requires extreme sailing. The Drake Passage is the chokepoint between South America and Antarctica, the place where the winds and the waves from the Atlantic and the Pacific meet and fight and roar and churn. A forty-foot swell is not uncommon; waves twice that are possible. A clickbait headline plays up the hazard, reading "Here's what makes the Drake Passage one of the deadliest places on Earth." To beat the hurricane season in the Atlantic, the *Meteor* had left Boston in early summer, but since the seasons are reversed in the southern hemisphere, that meant they arrived in the Drake Passage at the start of the austral winter.

Conditions were brutal. The *Meteor* kept getting turned back. It took multiple tries until they forced a way through. Melville had left Boston at the end of May and did not arrive in San Francisco until

October, four and a half months later. Another ship paralleling their route took even longer to make the trip; there were reports that year of fifty days of bad weather in a row.

With all that wind, it must have been great for ocean birding, or so somebody like me would think today. Melville too noticed the wheeling birds and wrote in his notebook about "speckled haglets," a kind of seabird that marks one's proximity to Antarctica. Modern editors, misled by an error in the *Oxford English Dictionary*, gloss this bird as a small white seagull, the kittiwake. Those editors need to go on more pelagic trips, since the footnote is entirely wrong. First, kittiwakes don't occur south of the equator, and second, if you consult whaling logs and range maps, it becomes clear that "speckled haglet" refers to the cape pigeon or cape petrel, which is gull sized but more acrobatic, and in all plumages is a bird strikingly checkerboarded in black and white. They are not shy and might have followed the ship hoping for scraps. I like to picture Melville paring curls of pork rind over the railing, chumming in the seabirds. Besides cape petrels, Melville would have seen thousands of shearwaters as well, and seabirds called prions, and small mini-penguins called diving petrels, and the dark-backed, voracious kelp gull, and five or even six species of albatross. Always something to study—assuming you did not get washed overboard in the pitching seas.

This is the part now when I have to stand up and use both hands and my whole body, so I can show how steeply a ship lurches side to side in weather like this, and yet at the same time, how high and fast the southern petrels arc up and then wheel back down again, carving the wind and going past too fast to count.

Even in modern times it still is bad. Last year I interviewed a ship's officer who used to work on cruise lines in Antarctica. One time in the Drake Passage, the windows of the bridge were smashed out by a rogue wave. He had come off duty and was in a cabin one floor below. He knew the moment it happened since he felt the impact, and then half a minute later, saltwater started running down the walls. He took a flashlight and rushed upstairs. On the bridge they had lost power, steering, everything. After ten desperate minutes, the crew had jury-rigged alternate steering lines and had covered the broken windows in plywood, so the stormwater was no longer rushing into the bridge. At the debriefing the next day, the still-stunned crew worked out the math. The wave that hit them had been ninety feet tall.

You can see a poor wretch go overboard in the churning seas of the Drake Passage at the fifty-minute mark of the classic adventure movie, *Master and Commander*. Even though it was filmed in a wave tank, the scene is harrowing and completely convincing. A sailing ship was always a three-way collaboration between wind direction, swell direction, and the choreography of the sailors in the rigging, making adjustments to keep the sails full and orientated in the most advantageous way. "Back then," writes oceanographer Helen Czerski, "sailing required a very physical connection to the ship—pulling, adjusting, gathering, tidying—straining the human muscle which continuously resculpted the ship to meet the demands of nature." The Age of Sail was also the age of men aloft, day and night.

On Thursday, August 9, 1860, Melville's notebook opens with a survey of weather, recording "a gale of wind, with snow & hail & sleet." Melville's cursive here is slanted, tight, hurried. When excited, he defaults to double-wide em dashes with a space on each side, as if he is in a fencing match with Emily Dickinson. His notebooks are in the Houghton Library at Harvard now, and with time the black ink has foxed to brown, making the pages look as old as medieval ledgers.

After the polysyndetic weather list, the notebook goes on to record the sad facts of a seaman's death.

> Ray, a Nantucketer, about 25 years old, a good honest fellow (to judge from his face & demeanor during the passage) fell this morning about day-break from the main topsail yard to the deck, & striking head-foremost upon one of the spars was instantly killed. His chum, Macey (Fisher) of Nantucket, I found alone in the upper cabin sitting over the body—a harrowing spectacle. "I have lost my best friend", said he; and then "His mother will go crazy—she did not want to let him go, she feared something might happen."—It was in vain to wash the blood from his head—the body [bled] incessantly & up to the moment of burying; which was about one o'clock, and from the poop, in the interval between blinding squalls of sharp sleet. Tom read some lines from the prayer book—the plank was sloped, and—God help his mother.—During the brief ceremony, made still more trying from being under the lee of the reefed spanker

where the wind eddies so—all stood covered with Sou-
Westers or Russia caps and comforters, except Macey—
who stood bareheaded.—The Chief Mate imputes the
fall to the excess of clothing worn,—excess, not as re-
gards comfort—but activity aloft.—The ship's motion
very violent today.

God rest your soul, Ray from Nantucket. You died because you
stopped to put on your coat.

In the nineteenth century, if there was a fatality at sea, the dead
sailor would be shrouded in their hammock or an old sail, weighted
with lead shot, sanctified with words from the Book of Common
Prayer—that is the "book" Melville cites—and then the body was
slid overboard. Traditionally you sewed a shroud shut by starting at
the feet and ending at the face, with the final stitches joining canvas to flesh through the end of the victim's nose. That helped keep
the canvas from filling with air and floating back up to the surface.
It also verified that the body was, indeed, fully and completely deceased. "We commit this body to the deep," the captain would have
said, which was followed by everybody putting their caps back on
and going back to work. The Tom who officiated was the *Meteor*'s
captain and Melville's younger brother, Thomas; if it was not too
rough, Tom and Herman Melville played chess in the evenings.

The day after the funeral, Melville was still trying to process his
feelings. "But little sorrow to the crew," the journal records, since "all
goes on as usual." He was somebody who watched himself watching
the others. "I, too, read & think, & walk & eat & talk, as if nothing

had happened—as if I did not know that death is indeed the King of Terrors—when thus happening: when thus heart-breaking to a fond mother—the King of Terrors, not to the dying or the dead, but to the mourner—the mother.—Not so easily will his fate be washed out of her heart, as the blood from the deck."

3.

One does not go into Melville's pages—journal entries, poetry, or fiction—to come out the other side happy or reassured. He is not Walt Whitman in that way, or even didactic, high-toned Emerson. (Ezra Pound once said optimism was a hygienic beverage from Boston, invented by Emerson.) If held up side by side with his contemporaries, Melville is stranger and flintier, not hard in the sense of difficult word choice, but hard in the sense that he is the same type of obdurate ore one encounters in a meteorite, denser than earthly metal and more difficult to forge. If you found some on the beach, you could ruin a good knife trying to pry off a slice, and in the end, you might donate the fist of metal to the local museum as another troubling and inexplicable curiosity. As the poet Susan Howe says, "Herman Melville is not comforting. Emily Dickinson isn't either. Maybe their work is too hungry for comfort, or just too vivid for comfort."

He was death haunted, as were others from his era. Melville had watched his father die of delirium at age forty-nine, and of Melville's four children, his son Malcolm committed suicide, and his other son, Stanwix, died from tuberculosis. Of the remaining two children, third child Bessie never wed, and that left it to the fourth and final Melville child, Frances (usually called Fannie), to continue the line. She lived until 1938, long enough to see her father's reputation start to rise up again. His unfinished last work, *Billy Budd*, was published in 1924, though it took until the 1960s for it to get modern, professional editing.

Grief, death, failure, black moods—these things made Melville, Melville. And seeing the whaling paintings of J. M. W. Turner in London in 1849, if a retrospective at the Met is to be believed. And of course the pocket notebooks, and reading Shakespeare, and the sound and the fury waiting for him inside the King James Bible's pentameter. Hawthorne and Queequeg. Speckled haglets. "Scutch", a verb, what one does to flax. A willingness to upcycle words like "dismasted," which is said of Ahab after Moby Dick had taken his leg. This is the same Ahab who "lived in the world, as the last of the Grisly Bears lived in settled Missouri." Things in the Melville universe happen "bamboozlingly"; Starbuck inhabits a body whose "pure tight skin was an excellent fit." An innkeeper is named Coffin, and his bleak, half-collapsed hotel is described as palsied. As a writer, Melville wonders why use a positive if a negative can be twice as grim, and hence asks us to worry about "strange mummeries not unmeaningly blended with the black tragedy of the melancholy ship." That makes so little sense that it makes complete sense. What a not un-odd duck Mr. Melville must have been, to have evolved into such a strange, wild, queer (in all senses) writer. Poet Tom Sleigh: "Surely *Moby-Dick* is a prose poem."

4.

Patagonia turned out to be a non-place for me. I thought if I went to the places Melville had walked, that the air, the clouds, the stacks of wrecked kelp would tell me how to read him, and how, in some small way, to be a writer in his adjacency. I was very misguided. Without success I looked for Melville in the cemetery at Punta Arenas and in the men's room of the Shackleton Bar. There are shipwrecks you can visit in Patagonia, and I went there and took pictures of them. I took pictures of the Chilean flamingos and the herds of llama-like guanacos. I took pictures of the people taking pictures of the penguins. I hiked and whale-watched. I collected shells. I drank wine. And in all of that, *nothing*. No dead writers. No magic insight. After two weeks, my windburned lips cracked and peeled; I was a failure at even the most basic elements of self-care.

It is true though that I had always wanted to go to Patagonia, to see glaciers and rhea birds and to explore the enigma of another, later writer. Bruce Chatwin's book *In Patagonia* had made me want to become an author more than any other title I can think of. Somebody like John McPhee is not my model; he is admirable in his stately thoroughness and his ability to use dialogue like a bounce light in a studio portrait. Chatwin's prose in contrast relies on juxtapositions and dagger parries, and, as critics now know, relies too on half-fibbing near-truths, to skip the narrative forward like a flung stone bouncing off the lake. Language as come-on: the famous quote about Bruce Chatwin is that he was out to seduce everybody, and "it didn't matter if you were male, female, an ocelot or a tea cozy."

If so, so what? There are many writers I admire on the page—bad boy Lord Byron, the serially unfaithful Ernest Hemingway, alcoholic and wife-slandering John Cheever—who in real life I wouldn't let my worst enemy date. Read great authors, that's my advice, but never try to marry one.

What with kids and working and divorce and other books and more working, it took me longer to get to Patagonia than I intended, and in the end, I went not for Chatwin or Paul Theroux (*The Old Patagonian Express*), but in search for some trace of Herman Melville. I should have known I would not find him. Before that trip I had followed Melville to the Middle East too. I think I imagined him as some kind of slab tomb or stone effigy, and that if I laid blank paper across him and made a rubbing, some of his power would transfer to me. If nothing else, the paper would make a nice wall hanging: "Melville's Tomb, by Charles Hood." In my self-assigned apprenticeship, even my journals began to sound like him. Here is something from a Cairo notebook: "Kohl-dark eyes watching me today from out of doorways and under veils, eyes and hands asking for money, cigarettes, recognition, eyes always saying, *I am a person, pay me attention.*" Was that me or Melville writing that? Both of us and therefore neither.

Melville on the Patagonia trip wrote letters to his children but not to his wife. He seems lonely in his notebooks, sometimes heartbreakingly so. Melville in Patagonia was not in charge of the ship—his brother Tom was captain, not him—and yet if he tried to identify with the common sailors, he was separated from them by age and social status and his corrosive propensity for brooding introspection. The ship was just a proxy death box anyway. The timber hull and canvas sails meant nothing; the cargo was everything. Melville seems never to have reconciled himself to the brutal fact that in shipping, men are cheap and easily replaced, and the directive is to press on, press on, always to press on.

After that 1860 trip, Melville was done with hard water, hard landscapes. Something during the journey changed him. He got to California after five months of rough travel only to turn around and leave San Francisco after a week. To go home, he gave up on sailing ships. He took a steamship to Panama, crossed by land to the Caribbean, picked up a new ship, and was back in New York a month later. He never went to sea again.

5.

You don't read Melville; you endure him, one page at a time. When *Moby-Dick* finally ends on page 573 of the Newberry edition, everybody and everything is gone: Ahab, the whale, the crew, and the *Pequod* itself. Only the narrator Ishmael survives, floating on a coffin, unmolested by shark or frigatebird. After two days he is picked up by a whaling ship we have met previously, the *Rachel*, gridding a search pattern as it looks for its own lost crew. Contrary to what the text implies, at the end of the book Ishmael is not truly alone, since we are there too, we the readers, the ones who made it that far and, stunned, are trying to make sense of it all. We too have spent the night on the "soft and dirge-like main." We too watch the

Rachel's sails draw near, nearer, until the lookout spots Ishmael and therefore spots us.

By this point we know the truth. The actual ocean, the one on nautical charts or flattened into the pages of oceanography textbooks, goes down six miles in the Mariana Trench. At its deepest, there is nothing but stygian blackness and hull-crushing pressure. Yet that shallow deviation from sea level does not begin to contain the ocean's most horrifying depths. Much deeper than any mapped canyon are the depths of blackness that artists are able to write, draw, paint, sew, and sing into existence. Their oceans, like Melville's, contain infinitely denser pressure, infinitely darker nights, infinitely sadder endings than anything science can imagine or describe. Their water can break your heart, and there is nothing, *nothing* you can do to stop it.

GHOST SHIP

The plan was simple. One hundred people would meet in New Zealand and go by ship—go by ship *slowly*—from Auckland north to Japan. Fifteen miles an hour, one petrel at a time. According to the brochure, the lucky birders on board the *Lorelei* would combine island visits with at-sea transects intended to guarantee sightings of the planet's rarest seabirds. Some birds such as the "lava petrel" (name given in quotation marks since even the potential genus is disputed) would likely be new to science. Others species like the Bryan's shearwater, first described in 2011, are slightly better known, but only have world populations in the low hundreds. Extra servings of dessert would go to whichever keen-eyed soul first spotted Deraniyagala's beaked whale, known only from eleven carcasses, three at-sea sightings, and some dodgy acoustic recordings. We were going to be in the heart of its expected territory, or *maybe* we were going to be—hard to be certain, since the home range in reference books is a line of question marks straddling the equator.

Most but not everyone on board were birdwatchers at the highest, most obsessed level. At that time, nobody yet had seen ten thousand species of birds in the wild, but we had on board people who were in the high nine thousands. If we had hit a reef and had to evacuate into lifeboats, you can be sure they would have kept on calling out new species. If there were two such lifeboats, each evening they would have shouted back and forth, comparing their respective lists, arguing who had logged the most new sightings.

The trip was scheduled to last one full month. This was leisure travel at its most exotic; a single ticket cost so much you could sell the family car and still not cover the price of passage. Almost every birder I knew wanted a spot on the boat. The expedition ship, the *Lorelei*, was comfortable, even downright nice, but that didn't matter. The trip only happens once a year; the route was everything. Most of us would have boarded a tramp steamer to be able to go. The problem was that even the most fanatical listers can't easily be gone for a month from their real-world lives, or else by the time the arrangements can be made, that year's berths are sold out.

For me, given the twin problems of cost and work, I had to deploy a one-time-only secret weapon: *retirement*. If I could manage to disentangle myself from my academic career, I could access both time and money, since post-teaching, the tyranny of the semester grid would fall away. Parallel to that I was due a lump sum payout for accrued overtime. I had to resign to receive it; until then the money was locked in a vault, untouchable.

It was an easy decision. I hand-delivered my resignation to HR. The same day I stopped by the credit union to see about a bridge loan between the end of my contract and the start of my pension. Covid was supposed to be receding, lockdown was lifting, and world travel was revving back up. I called an agent in England and got a shared-room berth on the *Lorelei*. Airfares were arranged. Guidebooks were ordered.

For the next month, my home address would be sea level.

* * *

Picture a map of the Pacific. Dividing yesterday from tomorrow, the international dateline starts at the top of the map and threads the gap between Russia and Alaska like a thin wire trying to slip unnoticed between two pit bulls tugging at the ends of their chains. It then plunges in a straight line to the equator, where a boxy dogleg shifts the line right to herd Kiribati and Tuvalu back into the same time zone. Then the line returns to a straight course and rockets toward the South Pole, pausing only to curve around the wagging index finger of North Island, New Zealand.

That was where I crossed it on the flight from Los Angles to Auckland. I landed in the dark in the rain, with my body utterly flummoxed about what day it was supposed to be. I was exhausted and wide awake at the same time. Even my toenails ached.

Besides the official agenda, I had a second, more personal itinerary. Since our transit would by happenstance retrace the Allied advance during World War II, I would have a chance to experience the global span of the Pacific war, where my father spent most of his service in the US Navy during World War II. If I could make sense of the landscapes and seascapes of his primary experience, then maybe I could make sense of him too. I had always admired my

206 DOUBLE HYENAS AND LAZARUS BIRDS

father—tall, strong, faith-driven, hardworking—but I resented him as well, since when his PTSD flared up, he took it out on women and children first. He had died by the time I set out, leaving behind a tangled briar of feelings that some days felt like a thicket too dense to penetrate. Even in death, he circled me and I circled him, each of us wary of what the other was going to do. "Don't make any sudden moves," I felt him saying. "I am watching you."

Few Americans remember this, but Japan once held an immense empire. In 1941, Japan's line of control included parts of China, most of Southeast Asia and Indonesia, the island of New Guinea, and islands across the Pacific most of the way to Hawaii. Later they even reached the Aleutians. In order to reconquer so much territory, the Allies enacted the "island hopping" doctrine. They retook only the islands they needed for air bases and shipping—the main targets, the most coveted real estate—and bypassed the lesser choices, leaving those islands surrounded and isolated. The common metaphor was to let the skipped-over islands "die on the vine."

In this approach, one island could be controlled by the Americans, and across a narrow strait, the next one was still Japanese. In the movie *South Pacific* an American plane, trying to decide if the islands below are friendly or not, dips low. This sets off a volley of antiaircraft fire. The pilot dodges, climbing back higher. Unfazed by the flak, he turns to his passenger. "Yep," he says, chewing an unlit cigar. "They're unfriendly all right."

To take any given island, the formula was the same: US Navy ships bombarded the bejesus out of the landscape, and then the marines waded ashore, doing the rest by hand. The 2010 mini-series *The Pacific* dramatizes this part of the war. It was almost always grim. The Pacific campaign saw the first combat use of napalm as well as the first widespread deployment of flamethrowers in infantry units. These close-in weapons were needed because often Japanese defenders survived the preinvasion bombardments, having had years to prepare bunkers, caves, and log-lined trenches. The ships lobbed immense shells for hours, even days, and the defenders waited it out. There then would be weeks of bloody, intense fighting. Since the Japanese called out "Medic!" in English to trick Americans into exposing themselves to snipers, on some battlefields if you needed help, the codeword was "Tallulah," as in the popular actress Tallulah Bankhead. This name was supposedly impossible for Japanese soldiers to imitate.

This all was personal. It was history, but it was also direct, intimate, personal. It mattered directly; it mattered day to day. My father's ghost was not the only ghost I brought on the trip. I had a cabin so full of ghosts there was hardly room for my travel pillow and camera bag. Besides my father's wartime service, I had grown up with exmarines who had fought in the Pacific, and their stories had slowly become my stories. One example out of many: During the fighting on Guadalcanal, out of ammunition and with his position overrun, one of the neighborhood dads fought an enemy soldier hand to hand. Grabbing the other man's mouth, the marine had ripped downward, tearing the flesh off the bottom of the other man's face, and then finished the poor wretch by disemboweling him with his Ka-Bar knife.

That wasn't the worst story. That war is hell was obvious to all of us, and at the same time there was the cheery *bang*! *splat*! *ka-ping*! of the comic book panels, the same panels that were later parodied by dotty Roy Lichtenstein. I hated the war and yet worshipped it, and in movies and action figures and even daily slang, it never left me alone, not even for one day. A simple task was a milk run, bombing slang for an easy mission. Hog callers were below-decks loudspeakers—"Now hear this!" Every night on our black-and-white set there were shows that turned the war from brutality into jolly adventure, programs like *Hogan's Heroes* and *McHale's Navy*. One night a war-glorifying, too-trivial episode of a show called *The Rat Patrol* triggered one my father's moods. I was eight, sitting on the floor, and in his rage, he threatened to kick my teeth in with his shoe. He hit me, my brother, and our mother. It was a small apartment; there was no place to go.

After his fit passed, he was his normal, congenial self. The next night we played Chinese checkers, the same as always.

I received the war stories in fragments, overhearing them by accident and often with the narrative reshuffled. Living inside me, they shaped me with envy and uncertainty, mocking my own ambition to live up to the implied masculine codes. Since then, I have gone alone in the jungle at night and rafted Class V water in Africa and solo-climbed frozen waterfalls without a rope, but even so, I still wonder if I would have the guts to take out an enemy machine gun nest by myself. It seems unlikely. Binge-watching *Band of Brothers* has not helped me to resolve my doubts.

* * *

The first day on the *Lorelei*, one very good thing happened. It seemed that I was going to have a two-bed cabin to myself. I had not paid for such luxury, and nobody told me about the upgrade, so for the first few hours on board, as we roamed the halls looking for our assigned quarters, I was confused and then annoyed. *Where was my roommate?* Was he in the bar? Had he twisted his ankle? What had happened to him?

Pretour, we had been issued firm instructions: *Don't miss the boat.* Given what this trip cost, each of us had every incentive to be on time. As we pulled away from the dock, I sat at the small room's small desk, starting to get angry. I could not unpack or claim a bed until we negotiated who preferred what. Nor could I go around the ship and search for him, since I didn't know what he looked like. He had to be a "he"—on nature tours, unless you're married, it is always a he bunked up with another he and a she with a she, no comingling of genders or undergarments—but after that, it was blank. I did not know his name, so in my head I had started calling the missing man Goober.

Whatever else he was, like me, Goober must have been a friendless oddling. Almost everybody on board had signed up with a buddy or spouse and did not need to be allocated a random roommate. Some people had even boarded in gangs of six or eight, or else had come with a name-brand birding tour. These birding groupies roomed together, ate together, prowled the decks together. At the recap session in the evening they all sat together, tour versus tour, little islands of self-sufficient humanity. Some of them even had matching T-shirts.

I too had hoped to have a companion organized for myself, a wingman to hold my place in line or to share my fears and triumphs, but ding, ding, ding, four friends in a row had dropped out, one after the other, until I had run out of candidates and had to risk luck of the draw. It was either that or pay for a private room, and while I was willing to spend the price of a car to sail on this trip, I was not willing to spend the price of two cars.

I admit that as a traveling companion, I am the best of roommates and the worst of roommates. While I try hard to be amiable and attempt to bury my snoring in the far side of the pillow, I am, on the inside, seething with nonstop opinions. On the outside, I feign allegiance to the *que será, será* school of philosophy. Inside my head, it is all judgment all the time. For example, to pick one trait of many, late people bug me. And one thing was obvious on day one: Goober was very, very late.

The idea of spending a month in semi-solitary confinement with somebody who was perpetually late made me anxious, and being anxious made me even more angry. Late people are the same kind of people who always lose their room keys, who leave wet towels on the floor, and who mismanage their alarm clocks. Late people borrow things—pens, razors, phone chargers—and never return them. You do not want to spend a week at sea in a tiny room with a late person, let alone an entire month. *Where was he?*

As I fumed, the bing-bong went off—the ship's PA system, named for its two-chime tone—followed by garbled instructions. I gathered it must be time to kit up in a life jacket and do the obligatory lifeboat drill. Oh great, Goober was going to miss that too.

As the drill ended and as Auckland fell further astern, I realized that whoever Goober had been in the abstract, in the reality of this particular journey, he was not going to be sharing my small-cabin universe. He had opted out of the trip and had not boarded, and who knew why. Perhaps he was tracking the worldwide rise of shipborne Covid, and that had warned him off. Perhaps he had read some of my books and wanted nothing to do with their author. Maybe he had been paying on the installment plan and his final check had bounced. Maybe he had simply overslept.

Whatever the reason, I was now emperor of an empire of one. *Woot.* What my subjects lacked in quantity they made up for in quality. I looked at the room with appraising eyes. It was mine, all mine, albeit in a sort of worn-out 1980s way that could have used new carpeting and a brighter overhead light. Two beds: one for me and one for my cameras. The binoculars were hanging from a hook by the door; nobody was going to knock into them, and the hook needed to support no other shirts or coats. My shoes were lined up in the closet. The towels, dry and tidy, were on the towel racks, arranged large to small. Reading book on the nightstand. Spare reading glasses in the drawer. If anything was not in the right place, I had only myself to blame.

This was all most pleasing. I wondered at the time if the ship's bad luck streak had been broken. Before this trip, the *Lorelei* had had her share of watery troubles. She had been bought and sold half a dozen times, with name changes and foreclosures and impoundments for debt. In 1996 the ship had run aground, and the passengers finally left for home on another ship. In 1997 it had run aground again, but near as I can tell, the passengers stayed on board for a few extra days, and the *Lorelei* was finally refloated. In 2013, it caught on fire. In 2020, it hit and sank a Venezuelan patrol boat.

All that seemed to be behind us. We were on our way, and we had left on time, and I had a private room. In celebration, I ordered a second glass of wine at dinner.

* * *

My father liked the routine and the camaraderie of military service, though when I asked him what he did in the war, he once replied, "I painted things." For navy ships to stay shipshape, they had to be scrubbed and painted constantly. His best day was when he was promoted to lookout duty, scanning for planes. Plum assignment since you were stationed outside, high up by the command bridge, and even on slow days there was always something interesting to look at.

He got this post by passing a test that measured how fast you could tell friend from foe. Unmarked silhouettes flashed on a screen, and you had to make a snap call—shoot it down or hold your fire. Some quiz planes were the shape of ours and others, nearly identical, were the shape of the enemy's.

Perfect score. He later claimed he did better than the others because the exam was on a Sunday. He himself was a teetotaler, while the other fellows (he said) were all too hungover to concentrate.

I doubt that explains it. My guess is that he was naturally good at spotting things, simply because when I was that same age and first took up birding, I too showed uncanny skill. In fact, he and I looked so much alike in our respective twenties we could have been twins, and so we must have shared many traits, eagle-eyed vision among them. When it came to processing visual input, his eye-to-brain wiring must have blazed with red-hot intensity, since he was ranked the top spotter on the ship.

Some distant ancestor had made a pact with the devil, or so I have to assume. When I was young my eyes were good enough to have been bionic, and once I could afford top-shelf binoculars, that was it: I could pivot in a slow circle and scour the landscape like the Eye of Sauron. Cataracts have stolen it all now, and as my acuity fades, I rely more on the autofocus of my cameras and on the mad skills of younger companions, like my friend José Gabriel. The problem was, José Gabriel was not along this time. He had other obligations, and the trip price was a deal killer. I had to find my own birds. Okay, fine: where were the big, slow ones?

The biggest, easiest-to-track ones were the boobies. We had three kinds: masked, brown, and red-footed. They circled the ship, taking swooping dives after the flying fish we were flushing, and at night, when it was too dark to hunt, they roosted on the ship itself. Boobies don't mind hanging out on manmade structures like navigation buoys or the crossarms of sailboat masts. Plywood or other debris is fine too. I once saw a masked booby in the Java Sea and its home was a Styrofoam picnic cooler, floating upright, stained with guano, back wall missing, and that cooler had become a kind of Roman chariot with an occupancy limit of one.

Boobies really liked our ship, circling around and coming in low on either side of the bow to chase fish and squid. In flight a booby is a cross between a pelican and a javelin, with a head that is large eyed and stout billed, the whole head and neck tapered and strong, lethal as a spear. They can dive thirty feet in the water on the initial plunge and then swim deeper from there, rowing with their long, pointed wings. Because they were large and obvious, they were easier to photograph than the shearwaters and storm-petrels. I really liked them.

Most days we also had visits from tropicbirds (hello, mentor Trudy!) and two kinds of frigatebirds. Let the others chase dickybirds in the thorny thickets and steamy jungles. I like my birds out in the open air, with a nice breeze and my bar tab open in the lounge. My camera liked the booby show as well. My pictures of them are among the best shots I got the whole trip.

* * *

The original itineraries—the ship's official one and my shadow one, reconciling memories of the war—were equally plausible, equally likely to succeed, or at least they were right up until the evening when Covid broke out on the ship. Overnight the *Lorelei* went from being an affluent expedition cruiser and the envy of equatorial waters to being a plague ship that no port would accept. Crew and passengers alike were trapped on board with no hope for parole. The journey had flipped from being the trip of a lifetime into an improv version of *Waiting for Godot*.

Luckily, not all of us were ill. I was fine myself, since as a people hater and as a person in a single room I had not had enough close contact with humanity to receive a triggering dose. Others on board, through potent vaccines or cautious misanthropy, had stayed well too. Each night at the recap of the day (when the day's bird sightings were shared, among other summaries), the unafflicted among us hoped to get on with the original trip, and we awaited news of some port, any port, that would let us come ashore, even just a few of us at a time in a well-sanitized zodiac.

Always the answer was the same: not here, not today.

We non-ill passengers muttered among ourselves, wondering and scheming. Couldn't bribes be paid, or couldn't the sickest people be cast away in lifeboats? We seemed to have an excess of sick people and lifeboats; let's shed some of both and see if that didn't up our odds. If the bridge staff had a plan, it was not shared with the passengers. For unknown reasons, the ship's doctor had been spirited away in the middle of the night; his replacement, a man who suffered from intense social anxiety, stammered a few words of greeting, then disappeared into his cabin. He did not emerge for the rest of the trip.

Trapped, I started to observe the humans as well as the birds. I had a vision of all of us becoming dirtier and hungrier as month after month passed, the captain steering in circles until the fuel ran out, and we finally would be forced to beach the ship on an uninhabited island. There, à la *Lord of the Flies*, we would devolve into feral warfare. I already knew the people I would eat first—not the plumpest guests, but the most annoying. My candidate list grew longer each day.

The staff I would spare. They were only trying to help. My favorite waiter was William. He was handsome in a serious, Jesuit kind of way, as if in another life he had helped to run a monastery in Spain.

After only a few days he knew me better than a spouse: he knew where I wanted to sit, how quickly I wanted to order (*very* quickly, always), which meals I had wine at and which a Diet Coke, two lemons, extra ice. I had my preferred morning and evening table in the regular dining room—a two-person table, as far from others as possible—though at lunch, when William was assigned to the aft bistro, it was trickier, since others might claim my favorite spot there before he could shunt them aside.

Sometimes at night, this or that intruder joined me at my dinner table. William and I might make quick eye contact then, with me signaling, "Can you believe my bad luck?" He would nod in sympathy. I did cheat on William sometimes, going not to the bistro for lunch but stealing silently into the main dining room. I did it only if the posted desserts were so especially good that I could not pass them up. One day I went to both venues in one sitting, first the dining room and then scurrying aft to meet William for a follow-up lunch at the bistro. Was it gluttony or guilt that made me feel so nauseous?

The waitstaff worked hard to make this floating plague ship as pleasant as possible, as did the cabin stewards. Officers had nicer uniforms; sometimes they came to the back rail to smoke. I watched them all, wondering where I fit in. If this boat never docked again, what job could I do? Among the crew, I didn't aspire to be the captain—too much responsibility, and what happened when we ran out of provisions and needed to start chucking people overboard? He faced hard decisions. Instead, I wanted to be one of the zodiac drivers.

They wore many hats, literally and symbolically, since they were also evening lecturers and shoreside hiking guides and general, all-around, on-call helpmates. They knew about the birds and plants and whales, and they spoke softly but passionately in a mix of Kiwi, South African, and Baltic accents. If the cliche about Antarctic employment is that even the janitors have Ph.D.'s, then on expedition ships, the zodiac drivers all have master's degrees in conservation with a minor in mechanical engineering and a certificate in polar studies.

Stoic, competent, uncomplaining: forget the navy. I had found my new role models.

* * *

Where does revenge fit into the story of the Pacific? It seems overly simplistic to say that the United States entered the war because Japan bombed Pearl Harbor on December 7, 1941, and that we fought until we got even at Hiroshima and Nagasaki, on the sixth and ninth of August 1945. Once revenge was enacted, we could dust off our hands and say, "Okay, that was that. We're even. The war can end now." It seems simplistic, and yet it may not be wrong.

At Pearl Harbor, one of my father's cousins, Wesley Whitaker Hood, watched his own brother, Joseph Ernest Hood, die on board the battleship USS *West Virginia*. My father was not there—he was chasing U-boats in the South Atlantic at the time—but these two young men, as close to him as his own brothers, were on adjacent ships side by side on the Sunday morning that will live in infamy. One died; the other survived. In 1945, was the surviving sailor,

Wesley, glad to hear about the atomic bomb? If he was like my father, then yes. We do not know how to act, as a culture—we are equal parts Old Testament (eye for an eye) and New Testament (love all; serve all, to quote the House of Blues). "Take that," we say to things that vex us, from mosquitoes to mice to the people of other nations, other faiths. Take *that*.

After Pearl Harbor, America rounded up 120,000 Japanese Americans and put them in concentration camps. Take that. It did not make America safer. It did not speed up the defeat of Japan. But it did make a nervous, angry, racist country feel better.

Take *that*.

I will give my father credit. I never once remember him referring to the people of Japan or Japanese ancestry in nasty terms. He never once used a slur in my hearing. He never even said "Kraut," and unlike the rest of his family from the South, he never once used the N-word. If he spoke about Mexicans, such as the people he worked with as a delivery driver, it was only because he knew the individual had been born in Mexico, and that was the culture they identified with. For him, "Mexican" was specific, not generic, and was never pejorative.

I will count that as a lesson he taught me without intending to.

* * *

Though we skipped the island-hopping that would have offered views of nesting birds and their behavior, and that meant we missed a lot of history too, there were still plenty of birds aloft and afloat. If you're a certain kind of birder, this next list will push your envy needle over into the deepest green. Here are the *Pterodroma* petrels—the seabird group also called gadfly petrels—that we saw during the journey:

Beck's petrel
black petrel
black-winged petrel
Bonin petrel
Cook's petrel
Gould's petrel
gray-faced petrel
Kermadec petrel
"lava petrel"
magnificent petrel
providence petrel
Tahiti petrel
white-necked petrel

To explain this for the nonfanatics, on a typical day trip in North America, either from the Outer Banks of North Carolina or from any West Coast port, even on the longest, most extreme pelagic trips, birders only cross paths with one species of gadfly petrel all day. Maybe if the expedition's luck is running red hot, there could be two species total. We had a baker's dozen on this trip, and they included the rarest, least known types. In the case of Cook's petrel, I have even held it soft and warm in my hands, when one became disoriented and landed on the deck at night. It was calmed, crated, and released in the morning.

In an attempt to be more centered and happier, I have stopped keeping both my mammal list and my bird list, stopped all my lists local and international, all my lists daily or eternal. I am clean and sober for going on three years now. First birds, then mammals: I quit both. When it comes to tallying new ticks, I do not care at all, not one eeny little whit, and yet—

—and yet I do still sort of care, and I care more than I let on.

I am not proud of it. I do not admire listers at all—one wants to shout, "Get a life!" at all the folks glued to the bow looking, yearning, for just one more treasure to add to the trove—and yet I am the worst kind of lister myself: a closeted one who denounces it on Sundays and still spends all his time thinking about it, wanting to do it, going to the dark web after hours to read about it in the middle of the night.

Most islands are good for lists since islands have species that have diverged from their mainland forms and speciated into something new, and if so, these new organisms are generally native only to that one place—they are endemics, in the language of science. And thus rarely seen elsewhere. The home base of the *Enola Gay*, Tinian Island, has an endemic flycatcher, the Tinian monarch, a bird I had ticked off my list on a previous trip before I swore off listing, other than a quick nightcap now and again.

At the same time, "only here" is risky in terms of long-term ecological success; if you want to be ubiquitous through the ages, better to be a generalist, the way humans are, or pigeons, or barnacles. The Tinian monarch for now is stable—the population survived World War II, which was the main threat—but could be in trouble if the invasive brown tree snake sneaks over from nearby Saipan or Guam, perhaps in a cargo plane or a shipping container. These snakes eat bird eggs and gobble down endemic lizards too. The BBC said this about them in 2022: "The American bullfrog and brown tree snake have collectively caused $16.3 [billion] in global damage since 1986. In addition to ecological harm, the invasive pair have ruined farm crops and triggered costly power outages." On the island of Guam, the Department of the Interior spends $4 million a year trying to control nonnative snakes. Guam has up to thirteen thousand snakes per square mile, and they short out power lines. Personally, I don't see how they can be eradicated, but maybe some superlethal, super-specific, antisnake zapper wand will be invented, and things can be rebalanced. As a footnote to this story, snakes first arrived there with US military supplies during World War II.

On this trip, endemics were a priority, and the main reason we kept begging to be allowed ashore. If we did receive permission to land, we rode in zodiac boats in the dark hours of early morning. Once on shore, let us say on an island like Kolombangara in the Solomon Islands chain, everybody heaped the life jackets on the beach. They could be sorted later. The nervous, laughing birdwatchers then climbed into flatbed pick-up trucks. They would be driven a thousand feet higher, up the slopes of a volcano. Dawn would be pinking the horizon by now. At a clearing in the forest, there would be a lookout. (Sooner or later, one always comes to a lookout.) The trip notes, written by one of the birding guides, explain how "from the lookout we had very distant views of a couple of [the endemic] pale mountain pigeons, with a flock of around fifteen birds later making a close flyby and one briefly perching in full view. The road leading to the lodge was excellent for birding, providing new birds such as the diminutive Finsch's pygmy-parrot, Solomons cuckoo-shrike, oriole whistler, and Solomons white-eye."

These same islands hosted endemic bats. Whenever I saw them, I was afizz with happiness and joyful dancing. *Which* bats? I am so glad you asked. Take the Temotu flying fox, handsome devil with a three-foot wingspan. Orange-brown fur. Black wings. Found only on four small map dots in the Solomons. I saw mine at a place called Nendo.

If I did still have an actively added-to mammal list, it would only enumerate the species that I have seen and that the top mammal fellow, the affable and generous Jon Hall, has not. This particular bat is now on that list—on the list I do not keep, which contains the species I do not count. There is no list, and so the bat cannot be on it, and yet, somehow, magically, it is.

* * *

As the ship motored on, we were never given updated information about how many among us were sick. The plague victims were confined to quarters, so we never saw them and so could not get an accurate tally. Their confinement had one daily exception. For an hour a day, all the sick people were taken to the saltwater pool on Deck 7, so they could walk around outside and receive a dose of medicinal sunshine. The pool was a safe zone since it was enclosed by dense glass walls. Their miasma could not spread to the rest of us—it rose up into the open sky, rising harmlessly to mingle with the frigatebirds and the cumulus clouds.

The sick peoples' transit from their cabins to the pool was treated so seriously you would think they were radioactive. The rest of us were bing-bonged back to our cabins to clear the hallways. Let our good air and the sickies' bad air never comingle. Once the ill ones were locked behind secure walls, we could reemerge and resume our business, so long as we did not try to access the pool deck itself. That was off limits.

This daily airing of the plague spreaders gave the more curious among us a chance for a sick person headcount, since from the top-most half-deck, a space called the monkey deck, if you crawled over the lashed-down lounge chairs and peered over a forbidden balcony, you could see almost all the pool deck in one view.

Knowing the count made one popular at dinner—"How many today?" Everybody leaned closer, trying to hear.

"Thirteen."

"Is that more or fewer than yesterday?"

"More."

Why we were never told the updated numbers, that I could not figure out.

* * *

Is war a comedy or tragedy? Don't answer, since it is obviously one of mankind's most singular tragedies. But we respond so often with comedy, the genres overlap. My father used to talk about riding an ox cart when he was on leave on Guam, perched high on top of a mound of sugarcane. He waved at the peasants like a raj riding an elephant. That moment would make a good scene in a movie, especially if he were being chased by military police on rickety bikes, or better yet, if the police were riding donkeys with funny saddles.

Somebody needs to knock over the fruit stall or end up falling in a giant mud puddle.

Even more reliable, any time you need a cheap laugh in a war movie, have the fellows put on a talent show in which men wear lipstick and grass skirts, coconut halves strapped to their chests like mermaid breasts. Funnier still if they can sing and do a choreographed dance number, as in the musical *South Pacific*, 1958. A year later, it got even spicier. Cary Grant and Tony Curtis, *Operation Petticoat*: it is early in the war, and there is not enough primer to cover their new, under-construction submarine, so they have to mix red with white, paint the whole boat pink.

Then they pick up some stranded nurses, and while under attack, to feign a direct hit, they ask the nurses for their help so they can eject bras and panties from the torpedo tubes. Fan yourself with a hanky, because the audience is left to imagine cramped quarters, sheer fabric, people pressing tight to pass in the corridor, the muscular sailors' thin, Marlon Brando T-shirts, and the nurses' unhindered breasts under their too-tight, button-challenged blouses. It all stays on the correct side of the Hays Code, but not by much.

We make different movies now. Among other things, leading men are even more handsome, women even more striking. Terrence Malick's version of *The Thin Red Line* is sort of a movie about the ground campaign on Guadalcanal and sort of a dreamy mix of AWOL idyls and small-town flashbacks. Everybody is gorgeous. The closest we get to an ugly person is Sean Penn; there is not an average joe in the entire film.

In my world, we don't want to be in movies, we want to make

them. I grew up in LA, going to church in Hollywood; I have a friend whose baseball cap reads "2.76:1," which is the aspect ratio of 70mm film. If you have to ask, you're not in the club. In my remake of the Cary Grant movie, products will be the stars: paint from Martha Stewart, lingerie by Agent Provocateur; an Aston Martin will pass along the seawall in the background. Sex it up: drag queens of course, and mix-ups in the shower. One gag in my version will be that while one crewman always quotes *Hamilton*, his buddy always quotes *Encanto*. (We don't talk about Bruno—we never talk about Bruno.) The plot twist will be that there has to be a sneak attack on an enemy harbor, only because of global warming, there has been a red tide event. Regular all-gray subs can't get through—they will be spotted—and so everything depends on Old Pink and its fey, Broadway-singing crew. They are almost caught a bunch of times, but in the end, they make it—just as we knew they would.

Probably more accurate to label them not war movies but battle movies. It has to do with duration and narrative arc. Critic David Thomson elaborates: "War is a malignancy in our nature and our society, the deep expression of our fear; while battle aspires to adventure and a thrill, like going to a movie, and trying to believe that we can handle fear."

My father hated all of them. Too shallow, too phony. In *The Fighting Seabees*, 1944, a tsunami of Japanese soldiers is about to overwhelm an American position. Navy hero John Wayne wires a bulldozer with explosives, setting it up to ram a petroleum storage tank. The plan works, sending a cascade of burning oil into the path of the Japanese, who retreat in panic, running right into an ambush of machine guns. John Wayne had no time for true military service since he was too busy saving America throughout the Pacific. In *Flying Leathernecks*, 1951, he is stationed on Guadalcanal, while in *Sands of Iwo Jima*, 1949, he shows up many places, including being pinned down on the beaches of Tarawa. Good work if you can get it: filming took place on Catalina Island, Leo Carrillo Beach in Malibu, and Camp Pendleton in San Diego. According to literary scholar (and decorated veteran) Paul Fussell, when John Wayne tried to appear before an auditorium of wounded marines—real ones, not movie extras—he was booed.

* * *

In the Solomon Islands, while the others birded on the slopes of Kolombangara, I kept squinting out to sea. Offshore was the Slot— the shipping channel between the islands that was a dueling ground between fast, well-armed Japanese navy ships—destroyers, whose name says it all—and the American sailors trying to stop those destroyers from getting through and delivering supplies. It was in the Slot where John F. Kennedy was almost killed.

In August 1943, Lt. Kennedy was in charge of a plywood speedboat called the *PT-109*. It had no armor, no defensive guns. Being small, fast, and nimble, a PT boat's job was to dart in close to a much larger Japanese ship, bang out a few torpedoes, and race off before being blown out of the water. He was doing this on a moonless night, with even the stars hidden by clouds.

Instead of success, one minute the PT boat was making a turn to avoid hitting the target, which was suddenly looming very large, and a few seconds later, there was a crash and an explosion, and the smaller boat was smashed in two.

Two of the PT boat's crew were killed; the burst fuel tanks covered the sea with burning gasoline; nobody at the main base knew where they were; no US ships saw the accident. In the middle of the ocean, surrounded by enemy-held islands, the survivors held on to the wreckage, which was slowly sinking.

The next week sees them swimming island to island, trying to survive without supplies and without catching the attention of the Japanese soldiers.

At the main base, their funerals are held; all were presumed killed in action.

In the end, Native islanders found them, and Kennedy carved an SOS note on the side of a coconut. They were on Naru Island, and the message said,

NAURO ISL
COMMANDER . . . NATIVE KNOWS
POS'IT . . . HE CAN PILOT . . . 11 ALIVE
NEED SMALL BOAT . . . KENNEDY

In the end, after a harrowing week, everybody had been rescued successfully. Kennedy was given medals, which he deserved. They may have helped him win the presidency.

I had heard about the *PT-109* all my life. In the middle of the night the *Lorelei* passed over the site where it had sunk. To me, growing up with stories of heroism and being old enough to remember

Kennedy's funeral—it is the very first thing I am certain that I *do* remember, in fact—this was a special place. At the right moment, I went out on the deck to pay my respects. I was alone. Nobody else came out, not that I expected that, nor had I invited anybody. I said a prayer in the dark and stood there, and finally went back to bed, feeling a bit foolish, a bit without a place in the universe. What good was I, and to whom?

That was one reason I envied bird listers their unrelenting sense of purpose. They were like mail carriers, undeterred by heat or snow or dark of night. Nothing got in the way of their next tick. The landscape was just scenery. For me, when I was on Kolombangara or Guadalcanal or the other places we passed by, I was often thinking about ruins and bunkers and the hulls of sunken ships. I was thinking about another name for the Slot, which was Iron Bottom Sound, so named because of the many dozens of ships, Japanese and Allied, that had been sunk in its waters. Supposedly the coral was rust red, end to end, from all the shipwrecks. Each day, on the land or back on board the *Lorelei*, I was there and not there; I was in my body, and yet I was dreamy and elsewhere. For the birders, Kolombangara was a green, happy place where one could hope to see Roviana rails and duchess lorikeets. War was not on their minds, except the war to see more birds than the fellow next to them.

It was off of Kolombangara, for example, where a lucky dawn sighting let us add Heinroth's shearwater to the *Lorelei*'s trip list, another species whose world population may be only a few hundred individuals. *Birds of the World* describes Heinroth's as "a very poorly known seabird whose breeding grounds are still yet to be definitely discovered." I am not sure what more to say about it, other than it lives in coastal waters of the Solomon Islands and I have seen one and you have not.

* * *

In my father's navy, breakfast was called shit on a shingle—that would be chipped beef on toast to the rest of us. Every war invents its own vocabulary. The Japanese called Guadalcanal "Starvation Island," and not without reason. "Bandits" and "bogeys" are words we still use today, but some older military terms are cringey cute, like "bubble dancing" for washing dishes by hand. (In my house, I bubble dance after every meal.) "Dog fat" as a word for butter is plain disgusting. A "thirteen-button salute" meant that one was ready for sex, named after the number of buttons on the front flap of navy dress trousers. Pearl Harbor itself as a destination was just "Pearl" for short, and the place name came initially from a Hawaiian word that referenced the pearl-bearing oysters once common in the harbor. "Put the Pearls back in Pearl Harbor"—that would be a good bumper sticker.

During this war, individual leaders mattered. Say "American Civil War" and right away somebody can answer, "Grant and Lee." The nerdiest among us can tell you the name of Lee's horse (Traveller) or where Stonewall Jackson's arm is buried (Virginia). My father's life was controlled by orders issued by Admiral Nimitz, who was the face of CINCPAC-CINCPOA. (Commander in Chief, Pacific Fleet/Commander in Chief, Pacific Ocean Areas). Forever wars like Iraq or Afghanistan no longer create the opportunities to center our attention on a single decisive person, male or female. Most of us can name more celebrity chefs or auteur directors than we can current leaders of the American military. In its long-ago-and-far-away-ness, World War II—and the movies we make about that period now—still let us believe in great leaders, the specific people whose vision and insight and courage could make a difference for one side or the other. Germany had Rommel ("the Desert Fox"); the American army had General Patton; Britain had Monty—Field Marshal Bernard Montgomery, viscount of Alamein.

Chester Nimitz was America's greatest admiral, one reason so many schools and bridges are named after him today. He was thin and active, ready to go on a speed walk at lunch or listen to classical music at night. He never pulled rank. After returning a salute, he extended a hand and introduced himself. "My name's Nimitz," he would say, as if people didn't already know it. He would meet with anybody who might have useful information, from the lowest officer of the smallest ship to the combined high command of the entire Pacific. Other top American commanders believed that a good leader ruled with an iron fist. Or else some people like MacArthur surrounded themselves with yes-men and flatterers. Admiral Nimitz wanted to be sure there were a lot of smart ideas in the room at once. Everybody should have a say, even when they disagreed. To dispel tension after his staff had been arguing, he told off-color stories, funny but never nasty. Let's all have a laugh, it's good for morale. And once done, let's get on with business. We have a war to win.

Japan's best-known leader was Isoroku Yamamoto, the architect of Pearl Harbor. He straddled multiple worlds. He played poker, spoke English, and in 1904 he lost two fingers during the Russo-Japanese War. He was descended from samurai. He had taken classes at Harvard. Although he masterminded the plan for a surprise attack at Pearl Harbor, he argued against ever enacting it, since he correctly predicted war with the United States would have horrific consequences for Japan. Having seen the auto plants of Detroit and the oil fields of Texas, he argued that the US would win any war that lasted longer than a few months. The country with the most factories was going to do best in a long, globe-spanning war—Yamamoto saw that very clearly.

Yamamoto's role in Pearl Harbor was well known, and he was so hated in the US that when the chance came to assassinate him, the attack was code-named Operation Vengeance.

Because the US had broken the Japanese codes, Nimitz and his staff knew that Yamamoto would be doing an inspection tour of the front lines on April 18, 1943. The Americans knew where Yamamoto would take off and where he was headed. They knew he would be in a two-engine bomber called a Betty, with his staff officers in a second plane, also a Betty. Six fighters, Mitsubishi Zeros, would flank them as escorts. The Japanese assumed they would be safe; no American planes ever visited the islands where Yamamoto was going. If a stray American reconnaissance plane turned up, the Zeros would take it out.

Revenge, revenge—should the Americans risk it? They could take out the architect of Pearl Harbor. To do so, they risked revealing their access to codes. On top of that, they might fail. And they might need him alive, later, to negotiate a surrender.

Aw screw it. Take that, Japan.

The day before his inspection trip, eighteen American P-38 twin-engine fighters were outfitted with long-range fuel tanks. If Yamamoto stuck to his schedule, intercepting him was simply a matter of math: If Yamamoto left from this one place and flew this fast, then at this time he would be at this other place. More math would put the attack planes at the same place at the same instant.

Each American plane was armed with a cannon powerful enough to blow up a tank, plus four .50 caliber machine guns. Each machine gun bullet was the length of your hand, fingers extended. A .50 caliber round can penetrate a half inch of steel plate. If Yamamoto stuck to his schedule, intercepting him was a matter of geometry and radio silence. Courage too, but success at that point would mostly be a simple application of industrial production: we had a lot of bullets, and all of them were reliable.

The American planes took off on time, flew at wave-top height to evade detection, and intercepted Yamamoto's flight near Bougainville in the Solomon Islands. Two American planes had dropped out after takeoff with engine trouble. That made it sixteen planes against eight: the main swarm of P-38s against the six Japanese fighters, and a hit squad of four planes against the two outdated bombers, one of which was unarmed.

The math and the bullets worked, and the Americans shot down both bombers within minutes. The staff plane crashed into the ocean, and Yamamoto's plane crashed into the jungle.

Bougainville at this time was still Japanese-held territory. It took two days for the Japanese search party to march through the bush to reach the wreckage. They discovered that Yamamoto had been hit twice by .50 caliber rounds, once through the shoulder and once through the head. Either shot would have been fatal. He had been thrown clear of the wreckage, and they found him under a tree, head down, with his hand on his sword.

I wish we had not done this. I am speaking selfishly, as I think about the impact his death had on me, my father, and our family. Taking out a foreign leader because of spite seems de rigueur—Seal Team 6 going after Osama bin Laden, or all the CIA poison cigar plots to do a hit on Fidel Castro. In at least the case of Yamamoto, the consequences became a morbid version of the butterfly effect. With Yamamoto gone, nobody was left who was going to be reasonable. We should not have done it; the United States would need him later. In war, it should always be business, never personal—we learned that in *The Godfather*.

At first neither side made his death public. The Japanese kept quiet for propaganda reasons as they tried to figure out how to spin the disaster, and the Americans did not want to tip their hand that they had broken the code. They wanted it to look like a random piece of good or bad luck, depending which side you were on. Some P-38s happened to be in the area, how about that. War sure is a funny thing.

A month later, once the news was finally released in Japan, Yamamoto was given a state funeral. President Roosevelt, when asked about it by reporters, feigned surprise. Was Yamamoto dead? "Gosh!" he said. It was the first he had heard of it.

* * *

I have not been to the crash site on Bougainville Island, even though Yamamoto's plane is still there. It would be a morbid kind of tourism, admittedly. Other things got in the way of my explorations, such as the time my flight was diverted back to Hawaii by a typhoon. Another time, right before I was going to go, I broke my leg, and so I lost a few months of travel because of that. My analogy is that if you drive enough miles, then one day you will have a flat tire. Similarly, if you hike enough trails, sooner or later you will twist an ankle or come close—come *this* close—to stepping on a rattlesnake.

I have never been bitten by a rattlesnake, but I know other people who have. Some of my travel stories sound worse than they were at the time. I have had an AK-47 held to my head in Ethiopia and have been lost in a whiteout in Tibet. These things did happen, true. I have broken my ribs skiing and my wrists mountain biking. Really these are all incidental things, small things, just the normal wear and tear of spending time outdoors. It can happen to anyone: Ernest Hemingway walked away from two plane crashes in a row. Artist Peter Beard once had his pelvis crushed by an elephant. The director Werner Herzog has been shot while giving an interview and once saved Joaquin Phoenix from a car wreck. The car was upside down and leaking fuel, and Mr. Phoenix was trying to relight his bent cigarette, which Mr. Herzog wisely prevented him from doing.

* * *

When I was little, I used to sit with my father outside our second-floor apartment. My mother said he was not allowed to smoke inside, so he sat on the steps at night, smoking Camels, eye level with the tops of the palm trees as we listened to the passing freight trains.

This was casual, normal. You smoked around kids because you smoked everywhere else. He and I did not talk, but at least we were together, ready for whatever came next. He had been through a lot, I knew that, but he was, it seemed, glad to be there. We were happy together, side by side, two friends sharing a quiet evening, two men, two travelers. Two loyal companions. It was normal for him, too, since during the war, everybody smoked, even the crews on submarines and even while submerged, all hatches closed.

For most veterans, the habit stayed once they got stateside. Ads crowed, "More doctors smoke Camels than any other cigarette." Our father finally quit smoking when my brother was born, so they never sat on the steps in the dark. It was hard for my dad to quit, and I admire him for it. After the war, on my former campus, I am told there was tension between vets and the rule-makers. At the time, the college shared grounds with the high school, and since the high school was nonsmoking, the college students were banned from smoking as well.

The newly returned men—they were nearly all men—were not pleased by these restrictions. On the other hand, the vets, who had cars and money and worldly experience, were very happy to ask high school girls on dates. Many of the young women said yes. The future is now: let the baby boom begin.

* * *

Only recently have I made another connection, one that should have been obvious. That was how even after retirement, my mom and dad never went to Hawaii. Florida sure, and all the national parks, New England in fall, but never Hawaii, never Pearl Harbor. I assume my father did not want to face it, and who am I to judge? I can barely stand to go myself, and I am a full generation removed.

Pearl Harbor National Memorial has become the single-most visited site in Hawaii, with two million visitors a year. When I went there recently, I didn't even need to wait for an Uber to get back to the airport; shared ride cars were dropping people off so often, the cars that had newly arrived barely had time to let one group of passengers out before the next group was ready to hop in. Pearl Harbor has become a tourist factory run on the conveyor belt principle.

On the original day of the attack, Hawaii was so obscure most Americans could not have found it on the map. It was not even a state then—it joined the other forty-eight states when Alaska did, in 1959. We speak of the Bronze Age or the cultures that invented Clovis point weapons. Perhaps I belong to the Formica People. Let our daybeds be filled with the downiest of Dacron. Add DDT to taste and stir vigorously. Strange now to think about it, but I am as old as Barbie, Jiffy Pop, the birth control pill, and the state of Hawaii.

* * *

We are each raised inside our own separate families, each with different parents from the ones our siblings know. All our parents lived in their own families as well, different from the ones we as children experienced. My family was not my brother's family, and his family, especially after I married and moved away, grew to be a very different one than mine. In the before times—in a prior, pre-brother family—things were hard because we were so poor. Dad was out of work, and my mother's mother tried to slip her a few dollars every week, without her own husband, my very controlling grandfather, finding out.

I remember when I was maybe four or four and a half, old enough to feed myself but not old enough to start kindergarten, there was a time when my father held me face down in a bowl of cereal because I had not finished drinking all the milk at the bottom of the bowl.

"In this family," he said with his hands around my neck, "we—do—not—waste—food. Got it?" Yes sir. I got that lesson and a few others besides.

Once my brother was born, there was a change in mood, a change in direction. It was as if somebody cut loose the sandbags, and the balloon of our family could rise up into the clear air. My father got a steady job. Eventually, there was a house in the suburbs. Our own washing machine. A better TV. First one new car and then two. The war was nearly over.

Before that, the war came and went a lot more often. In fact, the war was like a drunk uncle who showed up almost every night and didn't want to go home.

* * *

As our boat of birders headed north, the Japanese authorities indicated that they might relent and allow the disease-tainted *Lorelei* to slip into port after all. Until then we had been firmly banned—we could not even go within *sight* of land—but it seemed that (a) we might arrive on time, and (b) we might be allowed to haul our bored, overfed bodies up a gangway and onto Japanese soil. From there, shuttles would take us to a city bus station, where we could catch regular busses or trains to the airports and car rental counters of our choosing.

Before that happened, the captain had negotiated permission for us to make a closeish approach to Torishima. This is a middle-of-the-sea uninhabited volcano, 375 miles south of Tokyo, a few miles across, and it is the site of another providence petrel kind of story.

As recently as the 1880s, there were millions of seabirds on Torishima. It was the home colony of the short-tailed albatross, huge and white. Adults have a yellow-rinsed head and a bubblegum-pink beak ending in a blue hook. Juveniles start chocolate brown and blotch their way to white. Immense wingspan: largest seabird outside of the Antipodes.

Two things happened here, one bad and the other worse. Either thing could have made the short-tailed albatross past tense permanently. The smaller bad thing is that their nesting island kept blowing up, since it was and is an active volcano. There were major eruptions in 1871, 1902, 1939, and 2002.

The much larger and badder bad thing was that feather collectors moved onto Torishima in the 1880s and began to kill the birds for their feathers and down, which were used to stuff quilts, mattresses, and pillows. As with the providence petrel on Norfolk Island, the albatrosses didn't fear things that approached them on land. The birds sat on their nests, patiently being killed one after another. A worker could cull two hundred albatrosses a day, and at peak production, there were three hundred workers living on the island. The math is inexorable. Nothing can withstand that. Once the feathers were harvested, the dead bodies were turned into fertilizer. Dead birds piled up so high the company had to build a railway, just to shuttle the carcasses. Clean numbers are hard to come by, but a fair guess puts it at five million albatrosses that were killed in the first twenty years. It may have been ten million.

In time, the short-tailed albatross became extinct. Greed wins again.

The Japanese military had radar outposts on Torishima during the war, and after the war, during the American occupation of Japan, the US replaced those with weather stations to warn against approaching cyclones. Because of the volcano, those eventually had to be abandoned. The last albatrosses had been seen in the 1930s, and in 1949, after one final offshore survey, an American ornithologist at the Allied HQ in Japan declared the species extinct. End of the line.

Or so everybody thought.

As the British felons saved by the providence petrels had learned to their lip-licking joy decades earlier, many of these long-winged and long-lived birds take years to mature. In their early, nonmating years, these albatrosses stay out at sea. So even if you kill all the breeding birds and pluck and pulverize every last gawky, gooney one, you're bound to miss a few. That is because they only come to the nesting grounds once they are sexually mature, sometimes as early as age five, but more typically age eight or nine. Between fledging and return, they ride the winds from the Aleutians to Hawaii. So you can have a ten-year-old albatross that has not been back to Torishima since it fledged off the nest, which means it could have stayed safely at sea during all of World War II and even through the start of the postwar period.

That is exactly how it happened. In 1951, the short-tailed albatross was discovered to be alive after all. Back from the dead, and let's all do a heel-clicking jig. Besides the at-sea juvies, some adult breeders also may have been overlooked on the steepest cliffs. Combined, there were about fifty still-alive, reports-of-our-demise-have-been-exaggerated short-tailed albatrosses. With only one left, it is a sad end to a hopeless cause, as with Martha, the last passenger pigeon, or with a zoo's final thylacine. With two or even eight or ten, the same. But with fifty surviving birds, that is different. With fifty you have hope. With fifty you have the start of a new colony.

By then conservation was ready to step in. The species received protected status in Japan and across the Pacific. A decoy system was figured out, to encourage breeding. More became more, year after year. They have started nesting at new sites too, such as Midway Atoll in the Hawaiian Islands. The current population adds up to 3,500 and rising.

As the boat drew within a few miles of Torishima, we took turns squinting through spotting scopes to try to make out the white dots on green hillsides. Were they sheep? Piles of trash? Cairns? No—wa-*hoo*—they were short-tailed albatrosses, sitting on nests, and there was an entire mountainside full of them.

The boat drew closer. The mountain grew larger. First one, another, then another—the albatrosses were flying past us now, some so close I would have used a wide-angle lens if I had one with me on deck. This species turns up on California pelagic trips once every happily-ever-after or so, with one seabird leader often announcing at the start of the day that if we were to see one, he would buy dinner for everybody on the boat.

Here on their home island, we almost had too many to watch at once. This summary of the moment comes from a Birdquest trip log: "With hundreds of both short-tailed and black-footed albatrosses swirling around us we were treated to a real show as [we] circumnavigated the island, managing to pick out a couple of eastern buzzards soaring near the coast. As we left, we tried some chumming to draw in the albatrosses. This was unsuccessful but we were all distracted by the humpback whale which began following the boat, staying with us for over ten minutes."

All true, except for the whales. Not one humpback, but three, and we had them in sight for almost an hour.

* * *

Suicide bombing—whether it happens in Palestine or Iraq or inside the London Underground—strikes most Americans as unfair, as an action that violates the laws of "normal" warfare. We feel that way these days, and most Americans felt that way in World War II, when kamikaze attacks began to happen in the Pacific. Yet the laws of war are social conventions only, tied to one's culture, and in any case, to die intentionally for one's home cause is a choice that predates the wars of even the Greeks and the Romans.

By the end of 1944, Japan was losing every battle, at sea and on land. Their best pilots had all been killed. Their planes (the few that were left) were outdated. US submarines were sinking every ship Japan tried to send out. They didn't have enough fuel. They didn't have enough food. With no ships, they didn't have any way of retrieving the soldiers trapped on islands as the US bypassed them, and they couldn't transport new troops to enact new offensives. Hoping to do anything to fight back, the military took the few planes they had left and the final vats of fuel, and they turned regular airplanes into guided missiles piloted by suicide bombers.

Kamikaze attacks would never work. That should have been clear to everybody. Even if you blew up one ship, so what? There were a thousand more after that, and a thousand more behind that first thousand. America was the greatest war-making machine in the history of the planet. Blowing up one single American ship, even a big one, would be like going to a thriving ant colony, picking out a single ant with a pair of tweezers, setting it aside, stepping on it, and then turning back to all the other hundred thousand remaining ants and saying, "Did you see that? What about them apples, huh?"

The remaining ants will scurry and swarm, hustle and sally, same as always.

The kamikaze attacks did not win the war or even delay the end by so much as a day, but they certainly had psychological impact. Among other results was unambiguous resolve on the American side. There was a sense that there was no reasoning with the Japanese— look at how they act. As soon as we had an atom bomb ready, by golly, America was going to use it. Our murderous intent was rational, after all.

In my father's case, after surviving U-boats early in the war, he had been transferred to the Pacific. In the Atlantic, every torpedo fired at him had missed; his ships were never hit. At first it looked like good fortune had followed him. In the new ocean, he experienced combat often, but the ships he was on always came through okay.

That luck changed on March 11, 1945, a Sunday, shortly after dark.

Earlier that day, twenty-four planes had left Japan with the intention of committing mass suicide by diving directly into US warships. The plane being used was called a Frances, and a Frances was a two-engine bomber that looked a lot like an American Marauder, a British Mosquito, or a German Ju 88. The difference between them was that the Frances had a janky airframe and unreliable engines. It wasn't a robust design, and everybody knew it, pilots included. But Japan's production capacity was so wiped out by American bombing they had to use anything they had on hand. Plans were made, orders given, and twenty-four planes were loaded with one bomb each and enough fuel for a one-way flight. The crews knew setting out that they were being sent to die for the Emperor, and in so doing that they were to take some of those fiendish, cannibalistic, fish-belly Americans with them. How bad were Americans? *They were demons from hell*—everybody said so.

Twenty-four planes took off, but soon that number became twenty-three and then twenty-two and then ten and then five. Nobody shot them down; they just broke down on the way from their bases to the target zone. Soon five planes became three and three became two. A few of the crews landed at Japanese-held bases; most died in the sea.

By then it was dark. The American ships did not expect an attack that night. Ulithi Atoll in the Caroline Islands was far enough away from the war zone to be safe, that was the agreed reality. My father's ship, the USS *Randolph*, was at anchor, refueling before going back to support the invasion of Okinawa. The best spotters were off duty, my father included, as were most of the gun crews. My father was one of the hundred men watching a movie on the hanger deck. They were sitting through *In Our Time*, 1944, in which a young woman traveling to Poland with her British boss meets a count, and they fall in love as World War II begins. Male lead was Paul Henreid, who previously had appeared as Victor Laszlo, the Resistance hero in *Casablanca*. (The two films also shared a screenwriter, Howard Koch.) The sailors watching this second-rate romance had no way to know that Japanese planes were approaching them at 340 mph.

Of these final two kamikazes, the first plane mistook a baseball diamond on land for the deck of a ship and aimed straight down. It hit the ground with a loud and tragic fireball. The plane's crew died instantly.

That left one Frances bomber still in the air. It was armed with a single 1,700-pound bomb, and seeing an Essex-class aircraft carrier tied up at the dock, it swung into a nose-down, accelerating dive and hit the stern of the *Randolph* below the flight deck. The entire back end of the ship exploded. The resulting fire was so hot that by the end of the night, the flames had warped thirty tons of steel.

My father did what was expected of him, which was to grab a hose and fight the fire. He only talked about it twice afterward. From what he said and from what I can glean from the battle logs, my guess is that at first he was in shock but probably doing okay, despite the noise and the alarms and the flames and the emergency lights, and despite not knowing if more kamikaze planes were going to follow the first one, and despite being angry at himself for not having been on lookout duty that night, and despite this being for him year five of what must have seemed like a war that would never end, despite all that he was doing okay, or I suspect that he was doing okay right up until he came across a burning, severed hand, a hand that he realized belonged to the body of his now-dead best friend.

That was the moment when the war finally pierced my father's heart.

It pierces mine too, writing about it.

I had gone to the Pacific looking for a path back to my father and his generation, and as the *Apocalypse Now* prologue so aptly says, for my sins, I was given one. I had brought a small library of reference books, and I had studied them night after night in my Goober-less cabin. I wanted better clarity, better reasons for what connected to what. The children of the mutineers on the *Bounty*, the prisoners on their many cliff-rounded islands, the lost marines and their trapped Japanese counterparts—so much stoic heroism on so many oceans and yet so much loss and hardship, often for no good purpose. Maybe I was a version of Herman Melville, who spent all those long months sailing around Patagonia to San Francisco, only to turn around and leave seven days later.

We expect sacrifice to mean something, or at least we do in looking back at it. The loss should never be random, never be without a good, clean result. When Tom Hanks dies at the end of *Saving Private Ryan*, at least we know that Matt Damon will get to go home afterward, and at least our side, the good side, the right side, will be one inch closer to winning the war. Reconnecting the dots of the war in the Pacific, learning the repeated human efforts to destroy each other as we all heaved on the waves and wind—it didn't bring any insight in the end, only sadness.

In this case, on the night of my father's kamikaze attack, the sacrifice of the Japanese plane and its crew changed nothing. It stepped on one random ant, and even then, it didn't smash it. On board the *Randolph*, repairs began immediately. Work crews rotated through twenty-four-hour shifts, and the ship was back in service in less than a month. It served out the rest of the war, all the way until September 2, 1945, when the armistice was signed. "Thank God for the atomic bomb," or so newspaper editorials agreed at the time, and those were also my father's exact words when I asked him about it fifty years later. If he had been in the war room with Nimitz and MacArthur and other brass, I am sure my father would have said to use all the atom bombs in the world if it could help end the war.

In the Pacific, tenacity was transforming into suicidal futility. Even after the death of Yamamoto, even after it was clear that the kamikaze campaign had failed, even after the fire-bombing of Tokyo killed one hundred thousand people in one night, even after appalling losses at Iwo Jima and Okinawa, even after all that, a defeated but defiant Japan fought on, pillbox by pillbox. Every fiercely fought island battle reinforced the idea in America that the mainland of Japan would be defended the same way. Better to bomb from the air than to risk invading on the ground—on this President Truman, his chief of staff George Marshall, and Secretary of War Henry Stimson unanimously agreed.

This equation was not entirely valid. How experienced soldiers act in the field does not predict how untrained civilians will act when their homes have burned down around them. They might resist ditch by ditch, or they might panic and flee. They might be too malnourished to fight. Among other differences between soldiers and civilians, Japan's population was being starved into submission because of Allied blockades and daily bombing. Terrain differed too; fields of rice and terraced pastures can be navigated by American tanks in ways that wave-lashed coral reefs and jungle mountains cannot.

Nobody cared about that. All the American public knew was that in every battle, the Japanese never surrendered. Military planners felt the same way. And so we dropped the atom bombs, aiming at Japanese cities that until then had been exempted from conventional raids. We needed unharmed cities so that the terrible power of the new weapons would be more legible. Kyoto had been on that atom bomb target list, but out of regard for international heritage was taken back off. Otherwise, almost any city would do. First there was Hiroshima, August 6. On August 9, Nagasaki had not even been the intended target; the lead plane, *Bockscar*, had been trying to bomb the nearby city of Kokura, but other cities in the area were already on fire, and there was too much smoke to see the ground. The pilots diverted to the next closest alternate, which was Nagasaki.

The two bombs did what they needed to do. Japan surrendered.

After the end of the war, my father's ship transited back through the Panama Canal and began running shuttles across the Atlantic.

Former German prisoners of war were convoyed back to Europe, and in their place, demobilized GIs came back home to America. It was called the Magic Carpet. The navy needed sailors to crew the ships that made these runs, so for most of the sailors, the war was not over. My father didn't get discharged until the end of 1947, two years after Hiroshima. Sitting down to his first Thanksgiving dinner in almost a decade, in the one photograph I have of him then, my father is on the periphery of a group shot.

The family is seated at the table; his own father, a pastor, is at the head.

They have just said grace.

My father is wearing a blue suit and has recently gotten a haircut. He has shaved. He does not look happy or unhappy.

Mostly he looks tired and a little dazed.

He is home; the war is over. Or it is almost over. Not yet. But soon, he is sure. *Soon.*

* * *

When the *Lorelei* was finally allowed to enter the port of Yokohama in Tokyo Bay, it meant we could each go home as planned. We had arrived on exactly the intended date. For those with nonrefundable plane tickets, myself included, this was great news. It was a lovely, bright spring day, and the morning smelled like fresh laundry. The literal air didn't, but I did: I had saved a final clean T-shirt to wear on my last day, something tucked into the very bottom of my bag, so the day smelled like my mother's favorite laundry soap, which is now my brand too. The shirt smelled like me. It smelled like home.

The birders sounded happy as they shouted out the final species of the trip.

"Black kite!"

"Swallow!"

"Vega gull, Vega gull, *there*—is everybody on it?"

There was also a heron and some hangdog cormorants, limp as wet laundry. A swarm of martins. Tree sparrow. Mallard. A pair of eastern spot-billed ducks.

Birders gonna bird, down to the final tick of the clock.

For me, I was done with it. My bag had been packed for hours. Good-bye, room. Good-bye, Goober. I patted my travel pouch tied around my neck, feeling for the stiff, reassuring rectangle of my passport. Soon I would need to be a person who was identified and identifiable. I was wearing a wristwatch for the first time in a month, and handy in my shirt pocket was my Global Entry card, so I could speed walk through immigration when I landed in Hawaii. I had my pen and journal. My phone was charged. My shoes were tied snugly and my shirt tucked in: always look your best when entering or exiting a foreign country.

I checked that my fleece coat was in the top of my bag; after the heat of the tropics, the air felt cool. If nothing else, you can always use a coat as a pillow if there's a delay at the airport. I put my binoculars back in their clamshell case; my birding hat slid into the pack next to them. *Until next time.*

Our fits of madness can pass; our runways can be given over to bunchgrass and purslane; our parents can harm us, and we can love them and forgive them and say to them, "What strange and deep wounds you must have been carrying, to have acted like that. What was it like? Tell me, talk to me."

We none of us have to continue being our old, bad selves—we can set those habits aside, like opening the closet door one morning and deciding, "You know, this coat is ugly and never fit right. I'm not going to wear it anymore!" Give it to charity. Maybe even throw it away. We all can change—I believe that with the fiercest of fierce convictions. I think it is fair and honest to admit that given enough time, even humans—even nasty, shortsighted, territorial, backstabbing *humans*—can figure out how to do the right thing.

Good-bye, war. It is time for you to be over.

Good-bye, Dad. You did your best. I am sorry about the night when the kamikaze plane hit your ship and killed so many men, your friend included, and I am sorry I do not remember that man's name, and I am sorry about another night, a few months later, when the pilot of an American P-38 misjudged a practice run, stalled, and crashed into the deck of your ship all over again, killing him and thirteen sailors. That must have been so traumatic for you and the other survivors. I hope you were off duty, someplace below deck and far away. I am sorry about all the stupid TV shows I made you watch, and I am sorry, when you were alive, we did not talk more, did not hang out. I would take up smoking now just to sit on the steps with you. Or vaping—that seems popular, we could do that. You do not know this, but when you died, the navy covered your casket with a flag, and in a final ceremony, it was folded up and given to me. I have it in the living room, in a triangular frame. I see it every morning when I open the blinds.

Good-bye, *Lorelei*. Thank you for not sinking. I like any boat that
makes it a habit to go an entire journey without sinking. Keep up
the good work.

Shake my hand, tomorrow. Nice to make your acquaintance. I
plan to visit you very soon, but flying home, I must first cross the
dateline, and when I do that, today will become yesterday, and yesterday will last for weeks and weeks and weeks, until finally I land
in California, where the concept of yesterday is outlawed, and everybody walks with a slightly forward tilt because they are so eager to
leave today behind and plunge headfirst into tomorrow.

Yokohama, good morning and hello. I think my dad was here
after the war. Perhaps you remember him—slim fellow, my height,
gray eyes? Blue shirt and white hat. No tattoos. He might have asked
if there was a church, so he could go to service. I know that you have
a noodle cup museum now and a botanical garden. I will have to
make time to see them during my next visit.

Kraaw, kraaw—hello, loud crows. Crows yet not crows; these are
not the American kind I am used to. One of the birders reminds me
of the name in English, "large-billed crow." This species combines a
raven's fire-axe beak with the modest body of a regular crow. It is the
best of both, other than they look front heavy and about to tip over.
According to the field guide, they use these stout beaks to eat road-
kill, nestlings, frogs, berries. They smash nuts with stones. They steal
food from vultures. I am packed and ready, and if somebody could
fly me home on a chariot pulled by crows, if they could fly me home
right now, I would send that person a thank-you card once a year for

the rest of my life. It would be a nice card, and I would mail it early, so it never arrived late.

Good-bye, water. Stop me if I have told this story before, but the first time I died, I was five. It was Easter weekend, clear and hot, and in the middle of a normal afternoon, I drowned in a swimming pool in Las Vegas. How I fell in, I don't remember, but once I accepted the fact that I was dead, I could enjoy the greeny-blue color of the light at the bottom of the pool. Hours passed. What urgent beast dragged me back to the surface? It was a large fish, I think, or a mermaid, or my tall, strong mother, or maybe it was the gravitational pull of the moon, turned up a few notches higher that day. Or sunlight itself brought me back to the surface, and the shimmery color of air creased by sunlight. Or maybe I just imagined that I was taken back up into the air, and I have been made out of water the entire time. We are not ashes to ashes; we are water to water, seamless and pure. Once more I come back to the ocean, to its birds and its boats, its storms and shipwrecks, its threats and its promises, because sea level has always been a lie, and the ocean is everywhere, inside and around each of us.

Hello, water. I am home.

Notes and Image Credits

page 43   "The Half-Life of Salt" chapter reuses the title from my 2002 Hiroshima book, *The Half-Life of Salt*. Part of that book takes place at the *Enola Gay*'s now-abandoned training base in Wendover, Utah. Photographer Mark Klett also has a Wendover-based book, *The Half-Life of History*. He and I shared the same location and the same aesthetic interests, and for access, we both were affiliated with the Center for Land Use Interpretation. My "Half-Life" title came first and still feels current, so I have borrowed it back.

page 203   *Lorelei* is the name of a fictitious ship that sinks off-camera in a John Cheever short story. All other names are factual, as are all the historical details.

page 213   As we went to press, ornithologists divided a single worldwide species, brown booby, into two. This happens increasingly often. In the same 2024 revision, the American Ornithological Society split the Audubon's shearwater into a complex of five separate species: sargasso shearwater, tropical shearwater, Bannerman's shearwater, Persian shearwater, and Boyd's shearwater. When it comes to bird names, "here today, gone tomorrow."

For their field time, technical advice, and/or wise reviewing, thank you to

Charles Anderson, Pamela Anderson, Susan Anderson, Rafael Armada, Sacha Barbato, Debbie Berne, Paul Carter, René Clark, Chris Collins, Matt Coolidge, John Eakin, Holly Faithfull, Jon Feenstra, Bill Fox, Jonathan Franzen, Kate Gale, Kimball Garrett, Mike Guista, Jon Hall, John Haubrich, Abbey Hood, Amber Hood, Fred Hood, Marek Jackowski, Mathew Jaffe, Frank Lambert, Michael Light, José Gabriel Martínez-Fonseca, Elizabeth McKenzie, Coleen Moloney, Christine Mugnolo, Dave Pereksta, Carolyn Purnell, Matt Rainbow, Peter Ryan, Catherine Segurson, Santi Tafarella, Andrew Tonkovich, Jann Vendetti, and Cal Yorke.

At Heyday Books, in alphabetical order, thank you to Archie, Emmerich, Gayle, Kalie, Marthine, Steve, and all the other merry pranksters.

Image Credits

Photographs by the author are from his collection, except for shells, vi (Ernst Haeckel); tropicbird, 20 (Daniele Occhiato, Alamy); banded sea snake, 56 (Images & Stories, Alamy); providence petrel, 96 (Toby Hudson, Creative Commons 2.5); chinstrap penguin, 116 (Godot13, Creative Commons 4.0); *Hunter with Cheetah*, 140 (Getty Museum, object 91.GG.53); detail from *The Life Line*, 164 (Philadelphia Museum of Art, the George W. Elkins Collection, E1924-4-15); tide pool, 174 (Dennis Frates, Alamy); and porthole, 257 (George Roux, *Twenty Thousand Leagues under the Sea*).

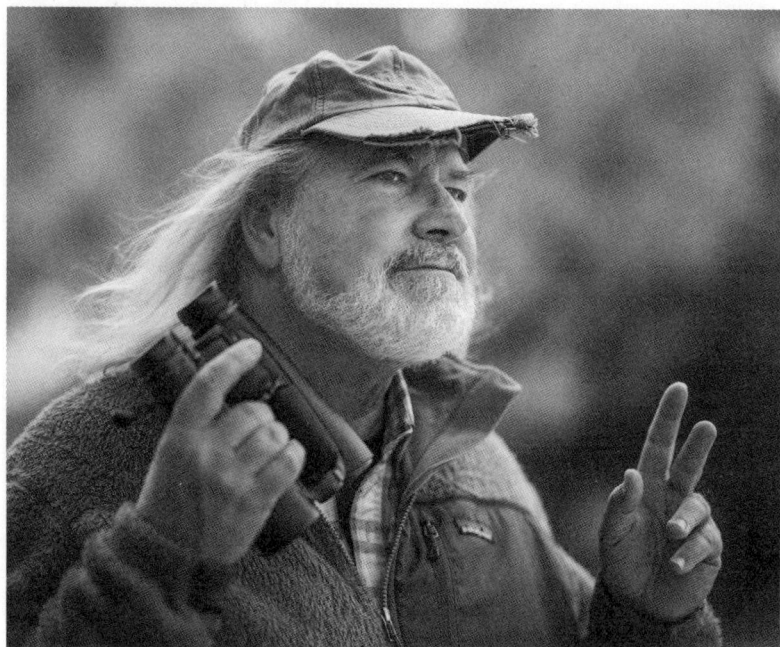

Photo credit: Killdeer Studios

About the Author

Poet, essayist, and water-skeptic Charles Hood has been a factory worker, a ski instructor, and a birding guide in Africa. His recent Heyday books include *Nocturnalia*, an appreciation of nature after dark, and the essay collection *A Salad Only the Devil Would Eat: The Joys of Ugly Nature*. Wildlife study has taken him around the world, from the high Arctic to the South Pole, and from Tibet to West Africa to the Amazon. Mammal 1,000 on his world animal list was a Crossley's dwarf lemur, Madagascar. (Mammal number 999 was a Malagasy white-bellied free-tailed bat.) Recently retired as professor emeritus, Charles lives in the Mojave Desert with two kayaks, two mountain bikes, two dogs, and five thousand books.

A Note on Type

The main text of this book is set in Adobe Garamond Pro. Garamond, created by Claude Garamond in the sixteenth century and reinterpreted for digital use by Robert Slimbach and Adobe, is one of the most widely praised typefaces in design history, noted for its exceptional elegance and readability. The headers are set in Mr. Eaves, designed by Zuzana Licko and published by the Berkeley-based foundry Emigre Fonts.